THE FULTON FISH MARKET

ARTS AND TRADITIONS OF THE TABLE

ARTS AND TRADITIONS OF THE TABLE:
PERSPECTIVES ON CULINARY HISTORY

Albert Sonnenfeld, Series Editor

For a complete list of titles, see page 281

THE FULTON FISH MARKET

A HISTORY

JONATHAN H. REES

Columbia University Press
New York

Columbia University Press wishes to express its appreciation for assistance given by Furthermore: a program of the J. M. Kaplan Fund in the publication of this book.

Columbia University Press
Publishers Since 1893
New York   Chichester, West Sussex
cup.columbia.edu

Copyright © 2023 Columbia University Press
Paperback edition, 2023
All rights reserved

Library of Congress Cataloging-in-Publication Data
Names: Rees, Jonathan, 1966– author.
Title: The Fulton Fish Market / Jonathan H. Rees.
Description: New York : Columbia University Press, [2023] | Series: Arts and traditions of the table: perspectives on culinary history | Includes index.
Identifiers: LCCN 2022015188 (print) | LCCN 2022015189 (ebook) |
ISBN 9780231202565 (hardback) | ISBN 9780231202572 (pbk.) |
ISBN 9780231554626 (ebook)
Subjects: LCSH: Fulton Fish Market (New York, N.Y.) | Fish trade—New York (State)—New York—History—19th century. | Markets—New York (State)—New York—History—19th century.
Classification: LCC HD9458.N5 R44 2023 (print) | LCC HD9458.N5 (ebook) |
DDC 338.3/72709747109034—dc23/eng/20220411
LC record available at https://lccn.loc.gov/2022015188
LC ebook record available at https://lccn.loc.gov/2022015189

Map, p. viii: The blocks as depicted here are based upon a map in R. H. Fiedler and J. H. Matthews, "Wholesale Trade in Fresh and Frozen Fishery Products and Related Marketing Considerations in New York City," Appendix VI, *Report of the U.S. Commissioner of Fisheries for 1925* (Washington, DC: Government Printing Office, 1926), 184. The exact size of the buildings built between these streets varied over time.

Cover design: Julia Kushnirsky
Cover image: painting © Naima Rauam
Printed and bound by CPI Group (UK) Ltd, Croydon, CR0 4YY

CONTENTS

Introduction: Between the City and the Sea   ix

1 FISH AND FISHING BEFORE FULTON MARKET
1

2 THE EARLY DAYS OF FULTON MARKET
15

3 FISH FROM FAR AWAY
30

4 THE HEYDAY OF NEW YORK'S OYSTER INDUSTRY
44

5 THE OPERATION OF A WHOLESALE FISH MARKET
58

6 FISHERIES AND THE FISH MARKET
75

7 TURTLE AND TERRAPIN
89

8 FREEZING, COLD STORAGE, AND IMPROVEMENTS IN TRANSPORTATION
103

9 FROM THE BROOKLYN BRIDGE TO THE FDR DRIVE
118

10 POLLUTION AND THE DECLINE OF NEW YORK'S OYSTER INDUSTRY
134

11 BUYERS
148

12 THE CULTURE OF THE FULTON FISH MARKET AND ORGANIZED CRIME
164

13 A MUSEUM AND TWO SHOPPING MALLS
179

14 RELOCATION
194

CONCLUSION: AFTER RELOCATION
209

Acknowledgments 223
A Note on Sources 227
Notes 233
Index 275

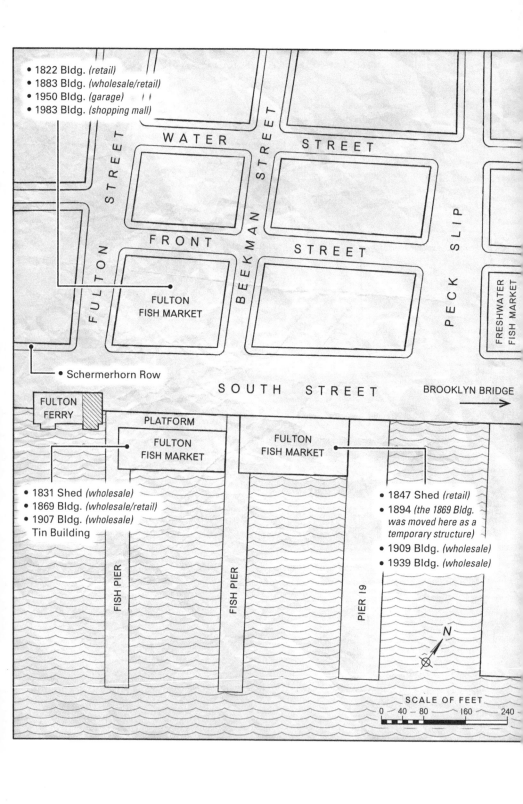

INTRODUCTION

Between the City and the Sea

Wherever there's water, you'll find fish, and luckily for us, most of them are good to eat. You don't need a lot of technology to catch and eat them. Under the right circumstances, you can even do it with your hands. The main limitation if you catch them yourself is that you are restricted to those that inhabit the water wherever you live. Since many fish migrate, you likely can't find all the varieties of fish that inhabit your local waters during every part of the year. However, because there are many different kinds of fish in the sea (and in lakes and rivers), a complicated system gradually emerged that allowed people from one place to eat fish that could only be caught in bodies of water located far away from them. Over time, different provisioning chains for every fish that's good to eat have come together to form a complex web.

These chains are largely invisible to consumers. We expect fish of all kinds to be available in the market, whether those fish swim anywhere near us or not. "For a long time I have been under the suspicion that Alaska and Florida are providing each other's storefront restaurants with bland frozen fish, in the way that some countries with cultural-exchange agreements provide each other with overly polite high-school students," joked Calvin Trillin in 1976.[1] This kind of exchange depends upon a recent invention called flash freezing, which allows fish to be preserved the moment they are caught. Thanks to this technology, Costa Di Mare, the seafood restaurant inside the Wynn Hotel in Las Vegas, gets fresh fish and shellfish like sea urchin from Sicily, flown in daily.[2] On the lower end of the price scale, Atlantic salmon farmed on Chile's Pacific coast is produced so efficiently that most Americans can afford it regularly, even if it too arrives by air.[3]

These might seem like particularly long supply chains, especially for a food so highly perishable. In fact, fish—especially expensive fish—have been among the best-traveled foods on earth for centuries, starting even before the onset of modern refrigeration. The commercial beef industry during the nineteenth century offers a good contrast for understanding how unique the long supply chains for fish of all kinds once were. Cattle fattened on the plains of Texas had to be run north to cow towns along the railroad lines. From there they would travel by rail to Chicago for processing in the South Side neighborhood known as Packingtown, but that only became possible during the 1880s after large packers like Gustavus Swift figured out how to use ice to keep the dressed meat they produced fresh all the way into East Coast markets.[4] Fish, however, often traveled similar distances over water decades before Swift created the modern beef provisioning system. Fish preservation methods like salting or drying date from ancient times. Keep a live fish in water and it will stay fresh until whenever you choose to kill it. Once ice refrigeration allowed America's railways to keep highly perishable food fresh, seafood transportation became big business. Pack a barrel with ice, put fresh seafood between the layers, and everything from oysters to herring to cod could find eager buyers across the country.

What Chicago became for meat, New York City was for fish, starting around the 1850s. Certainly, a city as big as New York might have the most fish eaters in it, but the presence of wholesale fish dealers mattered far more than the size of the consumer base. Fish wholesalers connected whole fleets of fishermen catching a wide array of species with the many grocers and restaurants that sold their product, both in New York City and for many decades, around the country. Concentrated in and around the city's famous Fulton Fish Market, these businesses both served and cultivated demand for seafood of all kinds. Unlike beef or pork, fish was (and largely remains) a wild food that can be taken by anyone with a rod, a line, and a hook. A business that specialized in fish had to regularize an inevitably uneven supply through a mixture of knowledge and technology. That's what separated the commercial fishing industry from sportsmen who saw fishing primarily as a means of recreation.

The original Fulton Fish Market, located alongside the East River just south of the Brooklyn Bridge, sat at the center of the fish provisioning system for much of the United States from the mid-nineteenth century to the mid-twentieth century. In the early days after it opened in 1822, merchants and peddlers sold foods of all kinds to a vibrant neighborhood. It was one of many general food markets around New York City.[5] Although seafood was sold there, fish dealers did not get their first permanent building until 1869. When the opening of the Brooklyn Bridge in 1883 lured away all the passengers coming to and from Brooklyn on the ferry and the port-related activities in New York moved from the East River to the Hudson River, the wholesale fish dealers completely displaced the other food sellers who had once dominated the area. It became an internationally recognized place for buying and selling seafood of all kinds. Much of the business of the fish market occurred out on the streets and sidewalks.

There were several structures officially designated as "the Fulton Fish Market" over its long life. This usually included more than one building at the same time. Generally, the City of New York owned the land and rented it to the fishmongers, who built and operated buildings on top. At its greatest size, the fish market's operations could best be described as six city blocks bounded by South and Water Streets on the east and west and Fulton and Dover Streets to the north and south. There were obviously many other places to buy fish in New York City, but this was the only place where you could buy them wholesale. In 2005, long past its prime, the market moved to a new, modern home at Hunts Point in the Bronx. It was the last of the city's markets from the nineteenth century (and earlier) to close down.[6]

The many supply chains running through Fulton Market began in countless fishing boats or nets and ended at the tables of restaurants and

FIGURE INT. 1 "Thursday Morning at Fulton Fish Market." During the late nineteenth century, Fulton Market was both a wholesale market that sold fish and a retail market that sold foods of all kinds. Hand-colored print by Al Hencke. *Harper's Weekly*, April 11, 1896.

South Street Seaport Museum, New York, NY

kitchens throughout the United States. After 1920, most of the fish that went through the market were consumed in the greater New York area, but their points of origin got farther and farther away as the technology to keep the seafood fresh improved. Fish is particularly perishable. Even the freshest, tastiest fish will suffer greatly when it is poorly handled. Although it can be mixed with other foods or covered in sauce, fresh fish is prized across cultures for good reason. Because of the limitations associated with preserving and transporting it, the creation of a vast web of effective provisioning chains to handle seafood of all kinds was a remarkable logistical achievement.

While the amount of fish that went through the market varied year by year, the general trend is clear from the available statistics. Fulton Market went from handling 31 million pounds of fish in 1881 to 394 million pounds in 1926. Then the volume gradually dropped throughout the rest of the twentieth century.[7] Between 1984 and 1987, it decreased 25 percent, to just 90 million tons.[8] Although Fulton was still the largest fish market in the United States, most of the product sold there stayed in the greater New York area because lots of people there ate fish. The overall decline in volume sold came as the result of new technologies of refrigeration and transportation that made it increasingly easier for suppliers to bypass the Fulton Fish Market entirely on the way from the sea to the consumer. Fulton Market still drew fish from everywhere because people in New York City ate an ever-increasing amount of seafood, but the fish provisioning system throughout the country and the world became increasingly decentralized. That system pioneered at Fulton Market contributed to the depletion of fishing stocks as soon as it appeared. As local shortages threatened supplies in Lower Manhattan, the wholesale dealers began to look farther afield. This approach magnified the impact of both this market and the market for particular fish.

That explains why understanding the impact of all the provisioning chains that intersected at Fulton Market requires examining multiple points on many of those chains and how information about the product traveled through them. The situation at the market depended upon how fish and seafood were caught, where they were processed, where and how they were stored, and where and sometimes even how they were eaten.

Every seafood provisioning chain starts in water, so pollution and water temperature affected their supply. Apart from oceans and rivers, other links in provisioning chains far away from New York City, like Gloucester, Massachusetts, and Baltimore, Maryland, affected the history of Fulton Market both as suppliers of seafood and as New York's rivals for domination of the wholesale fish industry. For these reasons, the history of the Fulton Fish Market takes in a wide swath of America, especially during the late-nineteenth and early twentieth centuries when the market was at the height of its influence.

On the other end of the chain, because the seafood traveling through the market went throughout the city and throughout the country, the history of the Fulton Fish Market is more than just a neighborhood story. At the beginning, the market had to be in the city to serve the city. The neighborhood around the market changed as the city changed. It began as a port, then became an industrial zone and finally a tourist district. These changes changed how the market operated. The history of the market takes in the history of restaurants and retail fish stores in the city too. By the 1860s, because of its sheer size, New York City held more people and organizations involved with some aspect of the acquisition, distribution, and sale of fish than anywhere else in the United States.[9] As boats gave way to trains, then trucks, the Fulton Fish Market depended more upon market forces and human activity than on nature itself. This ultimately led to displacement from the prime Manhattan real estate that it occupied. No single observer could take in the details of every product sold there, but the market became the place to go for people looking for expertise on both fish and seafood in general.

Everyone employed at the Fulton Fish Market was part of this larger system. In 1892, Alfred E. Smith began working at what was then known as Fulton Market. He was easily the most famous person ever to work there. He was an assistant bookkeeper for a firm called Feeney & Co., but in truth he served as a jack-of-all-trades. He packed fish in ice, rolled in barrels, and climbed onto the roof every morning with a telescope and reported to his bosses how much fish the boats headed toward the market had caught so that they could prepare to win more of it at auction against Feeney and Co.'s competitors.[10] If the boats lay low in the water, he knew they had a big catch. That meant lower prices. Knowing where each boat generally fished, he could guess which kinds of fish would be more expensive and trade for them before the price was set so that his firm could sell them later at a higher price.[11]

Like the fishermen he watched, Smith began his workday at four in the morning and ended it late in the afternoon, receiving twelve dollars per week.[12] His experience doing the kinds of hard industrial work at the Fulton Fish Market cemented his appeal with the voters who eventually made him a four-term Governor of New York and the (unsuccessful) Democratic presidential candidate in 1928. After Smith entered politics, he often compared his time at the Fulton Fish Market to the college experiences of his political peers. "That's where I got my formal education after leaving school," he once claimed during his time in the New York State Assembly. "I am the only member of this Assembly who can talk the fish language."[13] He often made a similar joke about his "cum laude" from the "F. F. M.," the Fulton Fish Market.[14] Indeed, a lot of knowledge about fish came to rest at the market because it sat in the center of the provisioning chain for almost every type of seafood available. This included knowledge from both fishermen, like when certain kinds of fish were most likely to become available, and buyers at the other end of the chain, like what kinds of fish consumers wanted to eat.

When reporters visited the market, they did their best to share the knowledge about fish that had accumulated there. "Down in Fulton Market," explained the *New York Evening Telegraph* in 1904, "every man is a piscatorial encyclopedia."[15] Everyone who wanted to make a living buying and selling fish had to be. There have always been an extraordinary number of different species of fish in the waters surrounding North America. Knowing which was which could be very difficult. For example, how do you tell the difference between a shad and an alewife? Then there is the problem of assessing quality. Fish is far more differentiated than any other food product. Besides the large number of species, there are large fish and small fish; fresh fish and not-so-fresh fish; fish that have been handled well on their journey to the market and fish that have been damaged. Understanding these differences was the kind of knowledge that kept firms at the Fulton Fish Market profitable.[16] Keeping that knowledge away from consumers was one way that less scrupulous wholesalers made their money.

Because employers and their employees often dressed alike, it was difficult to tell the owners from the salesmen or even the laborers. Reporters visited in order to inform their readers who couldn't come themselves or simply wouldn't visit anywhere as early in the morning as one had to arrive to observe most of the action. Naturally, since the media wanted to tell stories that people would read, the fish discussed in them tended to be the ones that most people ate. Journalists usually used the expertise of people working at the market as background to inform readers about fish in their reportorial voices. As was the case with so many other working-class Americans, the life experience of rank-and-file employees at the fish market seldom appeared in print. Even the salacious stories about organized crime in the market invariably focused on union bosses and their mob connections rather than the benefits of union membership.

The result was a lot of thick description of the Fulton Fish Market and its operation recorded on an almost yearly basis. Invariably, these stories highlighted the distant points of origin for much of the seafood available. Of all the knowledge that came to rest in the market, this may have been the most important, because knowing where to get fish that people wanted to eat is what kept the place in business as fish from local waters either disappeared or became inedible thanks to pollution. The changes in these points of origin over time reveal the impact of this industry on the environment around New York City and eventually the world, as more species were sourced from farther away as the techniques used to preserve them improved. When once obscure fishes became commonplace, their coverage in the media became less sensationalized as lower prices made them accessible to more consumers. At the same time, once common products like green turtle and terrapin gradually faded from descriptions as the market for them slowly disappeared.

The people who made successful livelihoods through the Fulton Fish Market encompass a large slice of the ethnicities present in New York City throughout its history. As the population of New York City changed, so did their taste in fish. For decades, retail fish stores that served every ethnic group in the city all bought their products at the Fulton Fish Market. Some of the people who ran them became wholesale fish dealers themselves. However, some immigrant groups were mostly excluded from the market except as buyers so that privileged workers could create opportunities for members of their immediate family—sons, uncles, cousins—to join the same business. This practice helped keep the market mostly white and overwhelmingly male for a very long time. For members of the ethnic groups who became part of the Fulton Market "family," working there offered unique opportunities for upward mobility because the fish trade was one of the few businesses at the turn of the twentieth century that was not dominated by a monopoly. Those people who didn't stay at the market often used their savings to start fish-related businesses of their own, such as buying a boat to become fishermen or starting a restaurant. In this way, the story of the fish market is the story of the American dream, even if limited to certain kinds of workers.

The success of the wholesale seafood dealers in the Fulton Fish Market depended upon the susceptibility of consumers to their knowledge and influence. During its early days, the wholesalers in Fulton Market had enough influence over consumer taste that they could create a national market for some species of seafood all by themselves. These products included sea scallops, pompano, grouper, and red snapper.[17] However, no fish would have pleased every consumer. That was the reason for keeping a variety available. Even then, different customers also cooked the same fish differently. The same species that might have ended up in fish stew during the nineteenth century might become ceviche during the twenty-first. Today, it is possible to get all kinds of seafood in New York City that that has not gone through any fish market at all because it is no longer necessary or even practical to coordinate all fish sales through a single location. This is the best explanation for the long decline of the market, which invariably made its many fans very sad.

"Every now and then, seeking to rid my mind of thoughts of death and doom, I get up early and go down to Fulton Fish Market," explained the legendary *New Yorker* writer Joseph Mitchell in 1952. "At that time, a little while before the trading begins, the stands in the sheds are heaped high and spilling over with forty to sixty kinds of finfish and shellfish from the East Coast, the West Coast, the Gulf Coast, and half a dozen foreign countries. The smoky riverbank dawn, the racket the fishmongers make, the seaweedy smell, and the sight of this plentifulness always give me a feeling of well-being and sometimes they elate me."[18] Mitchell often directed his attention to the disappearing working-class institutions of New York City, like dime museums and beefsteak dinners. He was a sentimental man who saw his writing as a form of historic preservation. Mitchell painstakingly described what these establishments were like—even if they were shadows of their former selves—so that others would understand what had been lost.

Mitchell first visited the fish market while working as a reporter for the *New York World-Telegram* during the early 1930s. He had always been interested in the fish business because it reminded him of the cotton and tobacco markets back in his native North Carolina. Reporting on the fish market gave him a window into an industry that few people understood as well as the Old New York that he wanted to romanticize.[19] After moving from the *World-Telegram* to the *New Yorker* in 1938, Mitchell wrote memorably about the Fulton Fish Market from the late 1930s through the late 1950s. Unlike other reporters given the assignment to describe all the activities on South Street, though, Mitchell was a people person. No matter what the story, he would invariably focus on a single individual and describe their life as a way of getting in touch with the place they lived or worked. Almost unique in the historical record, his stories are primarily about the men who worked around the market as well as in other parts of the supply chain that fed it.[20] While this technique is now widely associated with the *New Yorker,* Mitchell's profiles of men at various stages of the fish provisioning chain may be the quintessential examples of this approach.

Joseph Mitchell befriended people who worked in and for the sake of the Fulton Fish Market throughout his long life, but he did not write any profiles of market denizens for the *New Yorker.* Instead, Mitchell wrote about fishermen, a fishing boat captain, and the owner of the local restaurant Sloppy Louie's, which specialized in fish. In the years following his last original article for the *New Yorker,* Mitchell collected materials about the Fulton Fish Market in the hope of eventually writing an institutional history. Like so many of his projects from that long dry period of his life, it never came to fruition. He remained employed by the magazine until his death in 1996 but famously did not publish any new essays in any publication after 1964.[21] Nonetheless, even after his byline ceased to appear in the *New Yorker,* he continued to frequent the environs of the fish market.

Taken together, Mitchell's essays about fish and fishing echo the fish provisioning chain. His subjects worked at every link in the chain, from catching them to cooking them. Of all Mitchell's writing for the *New Yorker,* the essays with the closest connection to the operation of the market itself are a trilogy of fiction stories about Old Mr. Flood, the first two published in 1944 and the last one in 1945. The character Hugh G. Flood was a retired house-wrecking contractor, not a market employee. However, Mitchell based him upon "several old men who work or hang out in the Fulton Fish Market, or two who did in the past."[22] There is also more than a little of Mitchell himself thrown in for good measure. Mitchell's Old Mr. Flood stories gave words to a common attitude about the fish market that the author himself shared. The people Mitchell interviewed often lamented that the market had seen better days. The impulse of artists to depict aspects of the market in painting, photography, and other media reflects the same belief as Mitchell's: that its magical qualities had to be captured before it disappeared entirely.

Thanks in part to Mitchell, the Fulton Fish Market retained a reputation for never changing, except for being in a state of long, steady decline. "For the most part," wrote a reporter for the *New York Times* in 1985, "business is conducted much as it was when the market opened 150 years ago."[23] Around the same time, the South Street Seaport Museum, which includes preserving and interpreting the history of the market as part of its mission, endorsed the idea that "the Fish Market has changed very little since it opened on its present site 150 years ago."[24] This assumption is long-standing, and an obvious explanation for the decline that Mitchell lamented. That decline became a talking point for the people who spent decades trying to move the Fulton Fish Market to the Bronx.

This argument remains highly deceptive. The fact that the market held on to many traditions did not mean that it was frozen in time. The traditions of the market, like salesmen calling out the specifics of their agreements with buyers in code to bookkeepers nearby, never impeded its operation. Despite its many quirks, the Fulton Fish Market successfully adapted to changes at other points along long fish provisioning chains. It had more trouble adapting to decentralization because the cause of that change fell far outside the control of the wholesalers who ran it. A historical perspective on the entire life of the market and the system that surrounded it makes it possible to see all the changes at the Fulton Fish Market in the context of the industry it exemplified, as well as the contributions of this one vital supply point to the history of the American diet. Knowing the whole story also makes it possible to view this institution in light of the entire history of New York City. Changes in the market are best explained by considering the same trends that continually remade the entire city: urban development, shifting demographics, and especially technological change.

The most important change at the Fulton Fish Market was precisely in the way that firms did business there. Between 1822 and approximately 1907, it went from being an entirely retail market that sold foods of all kinds to an entirely wholesale market that sold only fish and other seafood. Very few of the reporters who visited seem to have understood that fundamental development, because so few of them ever mentioned it. Or perhaps they didn't notice this change because it was so gradual. Nevertheless, its effects were far-reaching. For example, wholesalers began to process many of the fish they sold right there at the market. Previously, this had mostly taken place either on fishing boats or in kitchens. Because most people who cooked fish hated the work that had to be done before they could even get started (skinning, boning, etc.), this led to the creation of a fish filleting industry that operated near the market, if not directly in it.

Changes in the buildings in and around the Fulton Fish Market were the most obvious physical manifestation of changes in the way that fish were sold there. To suggest that an institution primarily based in at least seven different buildings over the course of its history never changed doesn't take into account the activities of wholesale dealers who took over buildings throughout the neighborhood. The main market buildings, which housed stalls dedicated to both retail and wholesale operations, were rapidly run down by wear and tear or simply replaced because they no longer served the needs of the firms that occupied them. The neighborhood around the market changed drastically too, shifting from port-oriented to fish-oriented uses and then to mostly office buildings. Some of the oldest buildings associated with other commerce coming through the seaport were repurposed for the fish business and in many cases later abandoned. This kind of turnover explains why the structures in which fish were sold were never a particularly important aspect of Fulton Market. However, the development of what came to be called the South Street Seaport in the area surrounding the market during the late twentieth century greatly contributed to pressure that forced the wholesale fish dealers to move their operations to the Bronx.

The technology of fish provisioning changed too. Technologies that increased the efficiency of catching fish affected the market by increasing supply, decreasing price, and encouraging the wholesalers there to take on a greater volume of product. Changes in transportation, like the change from boats to trucks bringing the fish, made the entire supply chain more efficient. While local fish could be very cheap, getting fresh fish to parts of the country where they were rare required some very significant innovations. Thanks to gradual improvements in ice making and refrigeration, the fish that went through the Fulton Fish Market came from farther away than ever before. It also became possible to handle a greater volume of fish (without significant change to its taste) because technology helped wholesalers get their product out the door faster. Although the market itself lagged in the adoption of cold storage and freezing capacity throughout the twentieth century, the fact that wholesalers could still rely on improvements at other points in the chains contributed to the environmental disaster of overfishing worldwide. Yet despite the exploitation of refrigeration technology elsewhere, the preservation-related deficiencies in the market were one of the most important reasons that it eventually relocated to the Bronx.

Lastly, the kinds of fish the market handled changed frequently. This was the result of wholesalers establishing new provisioning chains that allowed them to bring in seafood from farther and farther away. In Fulton

Market's early years, the fish came from local waters. By the mid-nineteenth century, an increasing variety of fish came from anywhere between the waters off New England to those off North Carolina. As overfishing depleted supplies of popular species, wholesalers introduced new ones to their customers, often at higher prices. Some of the changeover was driven by wholesalers looking for new markets for whatever fish happened to be available. As the population of New York itself became more diverse, new immigrants wanted to eat different fish, so the Fulton Fish Market provided them. Retail store owners who bought at the market sold their fish in ethnic neighborhoods and catered to members of these groups. When fresh fish could travel on airplanes, it became possible to provide new immigrant groups with the exact same species they had consumed in their home country. In the market's final years, the wholesalers sold hundreds of different species caught in waters all over the world.

Starting with the concentration of the city's fish business in a single location and ending with the technological developments that decentralized the industry, the wholesale seafood dealers in the market adapted to all these changes with varying degrees of success. While some of the failures—most notably involving refrigeration and freezing—contributed to the false impression that the market constantly remained in a long, slow state of decay, considering the enormous transformation of fishing and the seafood industry since the early nineteenth century, it's kind of amazing that a fish market remained in Lower Manhattan as long as it did. This book covers both the evolution of this hub in an important food provisioning system in response to many different changing circumstances, as well as relationships that developed between the market and other links along individual chains for many different kinds of seafood. It also considers how changes in the urban landscape surrounding the market influenced the history of this landmark New York institution.

Although beef has obviously caught up to and surpassed where fish once stood in the American diet, it is worth appreciating the fact that complicated provisioning chains made the mass consumption of fish possible for the urban masses first. In 1893, a reporter for *Harper's Weekly* who visited Fulton Market observed that fish "was now the one rich food product within the purchasing power of even the man within the humblest of circumstances."[25] Higher volume, better quality, and lower prices explain why Americans ate more fish as it became more widely available to them. Fish have always been available to anyone who lived near a body of water, but it took centuries of effort for most of the bounty of the sea to reach all consumers, especially those who lived far inland. Understanding what has made fish something that Americans can and often do eat regularly is an excellent way to understand our ever-changing relationship with foods of all kinds.

THE FULTON FISH MARKET

1

FISH AND FISHING BEFORE FULTON MARKET

The fishing industry in and around New York City left little physical evidence for future historians to study. Fish are highly perishable, and once eaten, they are gone for good. Oysters are an exception to this general rule because their shells remain after the consumption of the creature within them. When the Lenape, the Native American tribe that preceded Europeans on Manhattan, and then the Dutch and English controlled the land that is now New York City, the waterways had immense oyster beds. Writing in 1740, the Swedish naturalist Peter Kalm observed, "The Indians, who inhabited the coasts before the arrival of the Europeans, have made oysters and other shellfish their chief food; and at present, when they come to salt water, where oysters are to be got, they are very active in catching them, and selling them in great quantities to other Indians, who live higher up in the country."[1] Evidence of this was visible on maps. In New York Harbor, Ellis Island was once called Oyster Island. Beadle's Island—now called Liberty Island—was once called Great Oyster Island.[2]

If this weren't enough proof, the Lenape also left their oyster shells in great heaps (which archeologists call "middens") that are still visible in the greater New York area because those shells don't decay. Archeologists have determined that some of them are thousands of years old.[3] It makes sense that the Lenape ate oysters and that they spent much of their lives close to large bodies of water. The sea and the large rivers that surrounded Manhattan provided human beings a steady source of easily acquired protein. Oyster beds were stationary, self-renewing, and often in water shallow enough that it could be waded into in order to acquire their bounty—offering some of the easiest food gathering possible for something that wasn't a plant. Since there was water all around Manhattanites during the precolonial and colonial eras, oysters had a very short supply line.

Of course, the Lenape and the early colonists also ate fish. When Europeans arrived, the Lenape sold them fish that they caught using both nets and spears. These included cod, bluefish, flounder, brook trout, and perch. The Lenape fashioned canoes, took them off the west side of Manhattan, and dropped lines into the water with hooks attached in order to get whatever fish was in season. Since they sat at a crossroads for so many different kinds of migrating fish, catching them at all times of the year would have been easy. Manhattan before European settlement was also criss-crossed by streams and dotted with shallow ponds.[4] Off of Staten Island, local tribes caught striped bass in large nets and dried them in the sun to provide food for the winter months.[5] With comparatively few humans compared to fish, all these practices were sustainable until Europeans disrupted this balance.

When the Dutch founded New Amsterdam, which would become New York, they chose its location because of the quality of the harbor. The fish then in New York Harbor were the most obvious manifestation of the bounty surrounding Manhattan Island. The Hudson River from Sandy Hook, New Jersey, to Troy, New York, was a spawning ground for many species. Tens of millions of fish from a wide range of species passed through the harbor as part of their annual migrations (and to some extent still do).[6] "Practically all the waters and rivers of [New Netherlands] abound in fish," wrote the Dutchman Adriaen Van Der Donck in 1655. He noted that the cod, who swam farther away from the mainland, "are plentiful. If people were to go in for it on the basis of experience that has been gained, shiploads could easily and cheaply be had nearby."[7] It is easy to tell that the idea of catching shiploads of cod was appealing to him. Cities depended upon agriculture to feed their inhabitants. Over the course of the history of New York, the source of most of the city's food moved farther and farther away from the city itself. However, as long as the waters around it remained relatively clean, there was always a source of protein very close to the center of the population. All you needed to take advantage of it was a pole or a net.

There was a market for fish in New York City long before there was a physical market in New York City devoted to fish. People could make a living as fishermen from the colony's earliest days, although Americans strongly preferred meat well into the nineteenth century.[8] Based on archeological digging through landfills, New Yorkers didn't eat much fish until 1760,[9] even though, as a 1753 chronicler of the city noted, "none of the colonies offers a fish market of such a plentiful variety as ours."[10] Shortly thereafter, the city's fish consumption spiked because the colonial government promoted ocean fish as a commodity, offering bounties to fishermen who caught and sold it. In 1773, the state legislature offered a bounty to encourage the consumption of an even greater variety of fish.[11] Nevertheless, that variety generally came from local waters until preserving fish with ice became common decades into the nineteenth century.

Fishermen could find more than fifty varieties of fish by venturing no farther than the waters off Long Island, Staten Island, and New Jersey. They returned to the city with their catch and sold it right off the pier at whatever market in the city could accommodate them, often pulling up to a nearby pier and selling straight off the boat. The Cortelyou family in King's County sold 25,000 shad each year as early as the 1790s. Staten Island fishermen began New York's oyster trade right after the turn of the nineteenth century, long before that industry would come to dominate the city's cuisine.[12] Even people with little capital but a small boat or a net could devote their time to fishing and benefit from the emerging market revolution, which made it increasingly likely that somebody would buy whatever they happened to catch.

Americans looked down on all kinds of fishing until well into the nineteenth century. The English writer Izaak Walton is widely credited with starting the pastime of recreational fishing with the publication of his book *The Compleat Angler* in 1653, but that book had almost no audience in the United States until it was republished in 1847.[13] Fishing for sport tended to occur inland on lakes and streams for freshwater fish. Commercial fishing on the ocean remained suspect. James Fenimore Cooper illustrated the difference between commercial fishing and traditional sport fishing in *The Pioneers*, the first book in his series of Leatherstocking Tales, originally published in 1823 but set during the 1790s. "These fish," the sport fisherman Marmaduke Temple explains to his daughter, "which thou seest lying in piles before thee, and which, by tomorrow evening, will be rejected food on the meanest table in Templeton, are of a quality and flavor that in other countries, would make them esteemed a luxury on the tables of princes and epicures." For Temple, the waste of quality fish—even by the poor—is a sign of the fecundity of his northern New York landscape. He thinks any kind of exploitation of this resource morally acceptable. This is not true of the hero of the book, the woodsman Natty Bumpo, who thinks it sinful and wasteful for anyone to catch more than they can eat.[14] This position eventually won out among sportsmen, but not with wholesale fish dealers.

The first modern fishmongers created a new link in the supply chain that made it possible to significantly increase the volume of fish for New York City and beyond. They began to deliberately fish in search of a surplus, then preserved and sold that surplus for a profit. They moved the fish they caught to where the customers were, selling to people who no longer had time to go fishing themselves. While fishing was hard work and didn't offer a lot of social prestige, it did offer the potential to become very rich, as the market they helped to create grew over time. Individual sport fishermen working with primitive technology could never make fish a viable commodity for a large population because they couldn't secure an adequate supply, let alone preserve it long enough to reach consumers who did not live anywhere near where it was caught.

Despite these limitations, the apparatus for obtaining, transporting, and preserving fish was big enough to severely deplete European fisheries as early as the late-medieval period.[15] Adam Smith in *The Wealth of Nations,* the gigantic 1776 tome that birthed the discipline of economics, argued that the most important facts about the fishing industry were its dependence upon geography and the huge amount of labor it required: "The quantity of fish that is brought to market ... is limited by the local situation of the country, by the proximity or distance of its different provinces from the sea, by the number of its lakes and rivers, and by what may be called the fertility or barrenness of those seas, lakes and rivers, as to this sort of rude produce."[16] This was an era when all fresh fish were local by definition.

You couldn't get fish if you lived near barren waters because the supply lines were so short. Even in a coastal country, it was still difficult to get fish if you didn't live near a fertile river or the coast. Although the sea was widely perceived as offering a limitless supply, that was not always the case in some parts of the ocean close to large populations that included many fishermen. Therefore, fishermen needed bigger vessels to travel farther to places that still had ample fish to catch. They also needed better methods to preserve them. Those vessels and methods of preservation both arrived during the nineteenth century.

Granted, people could process fish in order to make it keep longer. A fish begins to rot the moment it dies. The color of the skin changes, becoming dull and unattractive. Rigor mortis sets in, and its muscles stiffen.[17] Enzymes and bacteria that help fish underwater go to work immediately to bring about decay. Refrigeration can slow this process but not stop it. However, like beef, most fish undergo an aging process where the enzymes in muscles soften the tougher parts of the creature after death, which actually improves its flavor while making it more tender.[18] The quality of a fish depends in large part upon how it is handled from the moment it is caught.[19] Before ice, many fishermen tried to keep their catch alive on boats to prevent decay, and when dispatched at the market it was likely better handled than it would have been on a fishing boat. That fish was far more fresh when it hit the table after being killed closer to the point where it was consumed.

After death, fish dried in the sun last longer. Salted, they last longer still. Salting fish may have began as a preservation process before salting meat became a common practice. Fermenting the salt fish creates "garum," a fish sauce that began with the Romans. However, these ancient practices took place around the world. Smoking fish was particular popular among Native American tribes. Contrary to Smith's implication, there was a trade in preserved fish going back to ancient times that didn't follow his rules of geography or efficiency.[20] A hierarchy of fish developed even before Roman times. Groupers, mullets, and tuna were the most prestigious, expensive enough that only rich people could eat them, while smaller fishes like anchovies were consumed by the middling classes. Most importantly, many cultures looked down upon preserved fish compared to fresh fish in large part because fresh fish tastes best.[21]

The main difference between preserved and fresh fish was that all the traditional preservation methods that went back to ancient times changed the taste, often for the worse. But taste was not the only advantage held by fresh fish. Making preserved fish palatable enough to eat took more processing, and storing preserved fish required space. Fresh fish, straight off the boat, was invariably the preferred solution for urban market economies with people who wanted cheap protein and minimal effort, because under a capitalist regime they always had something else to do with their time. While keeping live fish in water was always a possibility, no fishing boat during Smith's time had the space to keep enough fish to travel long distances and make that product a common meal among the lower classes.

A market is an abstract concept indicating that there are buyers ready to purchase a particular product. Up until the early nineteenth century, most Americans were not part of the market economy. Self-sufficient farmers might have bartered their homemade products, but as long as they did not become wage workers, they did not have the cash to buy anything at markets. Cities, by virtue of the interdependence of their citizens, were where Americans first encountered modern markets because it is very hard to grow all your own food in an urban setting. New York

City was at the vanguard of this market revolution.[22] By the middle of the nineteenth century, nearly everyone in America had become part of the market economy.

A market is also a physical place where retail or wholesale business gets transacted. Every provisioning chain required a place for customers to buy the product being provided. Toward the beginning of his 1867 book, *The Market Assistant*, Thomas Farrington De Voe explains what going to the market was like for most New York residents early in that century.[23] "Some fifty years ago," he writes:

> it was the common custom for the thrifty "old New Yorker" when going to market, to start with the break of day, and carry along with him the large "market-basket," then considered a very necessary appendage for this occasion. His early visit gave him the desired opportunity to select the cuts of meat wanted from the best animals; to meet the farmer's choice productions, either poultry, vegetables, or fruit, and catch the lively, jumping fish, which, ten minutes before, were swimming in the fish cars.[24]

The markets of this era usually sold directly to consumers, which limited their variety and handicapped their growth. This changed when wholesalers began to buy products in bulk and distribute them for sale throughout the city and the country. New links in the provisioning chain emerged to handle an increasingly complicated system for making food available to consumers throughout the growing city.

New York City established its first public market in 1656. Authorities passed comprehensive regulations before the turn of the eighteenth century to increase citizens' access to food by keeping the markets open every day, but limited unlicensed peddlers who might cheat them and also might have lowered prices by providing competition to licensed merchants. Nevertheless, New York's markets were still rife with corruption.[25] De Voe counted more than forty markets over the entire history of New York City in 1862, but that included multiple markets on the same site.[26] Sales of both meat and fish occurred outside the market system, but licensed vendors harnessed the powers of the government to punish sellers and distributors who tried to get around municipal regulations.[27]

The need for healthy provisions and fair economic exchange have made municipal regulation a feature of markets around the world since civil society began. As urbanization progressed, city dwellers became farther separated from the food they needed to survive. Moving food from where it was produced—or in this case, caught—to where consumers could purchase it became a job unto itself. Since distant merchants were more likely to cheat their customers, this created a problem of trust. It became the state's responsibility to solve that problem by regulating the people producing and selling food, and that was easier to do when commerce was concentrated in centralized markets.[28] By the early nineteenth century, building public markets in urban settings had become one of the government's primary responsibilities. Municipal authorities regulated them so that the products would be affordable to all. Preventing anyone in the supply chain from causing artificial scarcities to drive up prices was an important way to do this.[29] Fish arrived in the city on an irregular basis because there was no way to be certain when they would be caught, and with no way to store them, their prices varied tremendously.

As the city grew, it became increasingly important to build new markets so that citizens could get food close to their homes and save time transporting it to their kitchens. The location of markets had to be convenient to both buyers and sellers.[30] Advocates for public markets believed that providing a site for food to be sold promoted people to work hard by giving an outlet for their labor, whether it was hunting, farming, or fishing. In this way public markets strengthened the dominant political value system of the United States during the Early Republic. As the number of markets expanded, so did the size of the older markets because they all had to serve a growing urban population. Fulton Market was one of the new markets built to provision foods of all kinds to the people who lived in the neighborhood surrounding it. Cities like New York also forced food sellers into public markets because they wanted to be able to guarantee that customers weren't being sold products that might harm their health.[31]

Before Fulton Market began to specialize in fish, every market near a pier included fish stalls or fishermen selling their product fresh off their boats. Catharine Market, the public market located directly north of where Fulton Market would be, had an area devoted to selling fish starting in 1805. Fishermen would simply pull up to the dock and sell right off the boat.[32] This was true at any New York market and for many fishermen. A French visitor at the turn of the nineteenth century counted sixty-three different varieties of fish available in the city's markets, along with oysters, lobsters, crabs, and crawfish. He wrote that the fish "are excellent, of all sorts, in great quantities, and at extremely reasonable prices."[33] This should be expected, as New York's waters remained a crossroads for fish of all kinds, as it had been in Henry Hudson's time. Over time, as the logistics needed to increase volume and maintain this variety grew more complicated, the advantages of channeling fish sales through Fulton Market led to the business growing there instead of at other markets in the city.

The problem facing fishermen who wanted to sell any fish was that they and the customers had to be in the same place at the same time, and every moment spent waiting for customers meant one less moment fishing. Middlemen (eventually known as wholesalers) became the link between fishermen and consumers. Fish cars were long wagon-sized containers full of water where fishermen could dump the live fish they brought in so that others could sell them. They were generally about twelve feet square and five feet deep, and they floated on the water like a raft. Deliberate cracks in the sides and bottoms allowed the water to flow in and out. A single fish car had enough space to keep 4,000 pounds of fish alive until workers at the market pulled them out in portions with a net for sale.[34] Fishing boats also had to dump their catch in fish cars because there was seldom enough space along the piers in markets to sell right off the docks. Fish cars made it possible to consolidate the catches of many boats, but they still took up a lot of valuable space—especially when they were lifted out of the water. The other solution to this problem was selling fish to middlemen who waited in a market for the people who would consume the product.

Fly Market, located near the East River at the southern tip of Manhattan, was founded in 1699. It had originally been called "Vly Market," from the Dutch for "in the marsh," since it had once been marshland. It was corrupted to "Fly Market" despite the undesirable association of flies and food.[35] Fly Market had been the city's largest public market during the eighteenth century and the very early nineteenth century, buoyed by its proximity to the recently established ferry that brought farmers into the city. It was also the site of a successful slave market. Other neighborhoods successfully petitioned for markets of their own, but none of them was ever as large as Fly Market.[36] It alone was responsible for two-thirds of the city's entire food trade.[37]

At Fly Market, fishermen clashed with authorities over their inability to sell fish off their boats because of crowded conditions. To conserve space, they had to pile empty fish cars on the wharf. Unfortunately, they needed them in the water at night when their boats arrived or their catch died before it could be safely transferred.[38] Catharine Market was the best retail fish market in the city before Fulton Market started specializing in that product a few decades later. Unfortunately, at the turn of the nineteenth century, other retailers attacked fish sellers for being a nuisance because of the smell. Besides the space taken by fish cars, fish dealers resented getting moved to the perimeter there despite their growing importance.[39] Yet nobody kicked the fishmongers out of the market because consumers wanted to be able to purchase fish.

By the end of the eighteenth century, Fly Market had become a dilapidated dump. Even though it had been renovated, it still attracted enormous opposition from citizens who found it lacking.[40] The Fly Market "was an unfit place," read one petition calling for its replacement, "being over a sewer, which in the summer-time is considered very unhealthy," and "it blocked up the street so much that mercantile business was partially stopped." That sewer had no cover in the part of the market that sold meat.[41] Indeed, the Fly Market also had swarms of insects surrounding it during the summer months. While a huge variety of fish was available there, the poor conditions under which that fish was sold became a serious drawback.[42]

The first efforts to close the Fly Market and replace it with a better facility began at the turn of the century. The area where Fulton Market would eventually be built had been river bottom during the late eighteenth century. William Beekman purchased the land in 1677; it was then far north of the city's center of population. The Beekman family expanded the tract by landfill, providing office and warehouse space for the city's burgeoning port.[43] The famous Commission Plan that laid out the city's street grid in 1811 planned out the entire neighborhood even while much of it remained water.[44] Water Street was the original waterline. Everything east of it is built upon landfill or pier. Peck Slip was a slip before 1817, when it became landfill. Many of the structures that still stand from this era have cracks in them because their builders couldn't wait for the land to settle before erecting the structures they needed to take advantage of the burgeoning trade at this important location during the early nineteenth century.[45]

The steam ferry between Manhattan and Brooklyn began to run in 1814, and it quickly settled on the area around Beekman Street as the most convenient docking point in Lower Manhattan. People coming in from Brooklyn didn't want to have to cross the peninsula to shop at Fly Market; they preferred a market convenient to their commute.[46] Passengers leaving the ferry were right at the front door of Fulton Market once it appeared in 1822. Schermerhorn Row, built in the same area and completed in 1812, included both merchants' offices and a warehouse. Both were evidence of a burgeoning commercial district that a market would only complement.[47]

The Common Council approved the building of the market in 1816.[48] The project was abandoned in 1820 when the money ran out thanks to the inability to negotiate compensation for landowners.[49] The Council reconsidered after most of the buildings were destroyed in a January 1821 fire that started in Sarah Smith's tavern and burned down between thirty and forty structures, mostly wooden houses.[50] As the *New York Gazette* explained, "The late conflagration has exposed to view this beautiful spot of ground, which has so universally been considered as the most eligible situation for a market . . . Its vicinity to the ferry, by which the marketing from Long Island will be saved the expense of cartage, a basin for smacks and fish-cars, unequaled in its purity, from its depth of water and the rapidity of its current, being decidedly in its favor for keeping and having good fresh fish."[51] Although Fulton Market was not planned exclusively as a fish market, fish were on everyone's mind from before it was even built.

The main opposition to moving Fly Market came from butchers who already owned stands there. They were concerned because their leases at Fly Market were supposed to be perpetual. The city ended up buying out their contracts.[52] Butchers also worried that if they built a bigger market elsewhere, they'd have more competition and their exclusive licenses bought from the city wouldn't be honored. One petition received in opposition suggested, "the Public Interest would be much better preserved by improving the Market on its present site, and by enlarging it by taking ground in its immediate vicinity." They also suggested that the sewer under the Fly Market might be sunk and arched over with substantial mason work, further improving the situation of those unhappy butchers.[53]

When the city first tried to sell the stalls in Fulton Market in the weeks before it opened, the butchers organized a boycott and a protest at the auction. As the Common Council reported, "The proceedings were riotous, persons were deterred from bidding in apprehension of violence and it was obvious that a combination was formed to obstruct the renting of the stalls." That violence consisted of the butchers, who were trying to organize a boycott of the bidding process, throwing a cigar dealer named Leonard into the river when he thwarted their plan by bidding on a stall. Under threats from the city, the butchers finally participated in the auction just days before the market opened in January 1822.[54]

Aside from the butchers, resistance to the building of Fulton Market was based on its expense. Built in just a year after the 1821 fire, the whole project cost a then-astronomical $216,768. Most of that money went to securing the land. But the city's costs also included $50,000 for filling lots underwater where the fish cars would eventually go.[55] This included not just the land under the building but also the land required to widen Beekman Street on the north side of the market so that it would be easier to access. Land speculators had bought up lots adjoining the space where the market would go after the initial charter for the project was made. When the city tried to avoid upgrading Beekman Street, the speculators sued. The New York Supreme Court sided with the speculators, arguing that the whole project was for the public good, so the burden should not fall upon people who had put faith in the first plan being carried out.[56]

Because of its proximity to the Brooklyn ferry, residents of Long Island were particularly fond of the location because they could shop on their way to and from Manhattan. "The new location being situated adding the (Fulton) ferry, will be to us a great convenience, remarked the *Long Island Star* in February 1821.[57] "Perhaps no site in our city combines so many advantages for the location of an extensive market, as the one between Fulton-slip and Crane-wharf," read an anonymous letter published in the *National Advocate*. "It is sufficiently spacious; it is in the

FIGURE 1.1 Fulton Market, 1828. "Fulton St. & Market," from the printmaker W. J. Bennett.

The Miriam and Ira D. Wallach Division of Art, Prints and Photographs: Print Collection, The New York Public Library. New York Public Library Digital Collections

immediate vicinity of our great ferry to Long-Island; and perhaps there is no spot on the East River so central for men of business. Our fish-market will then equal any in the U[nited] States; and, in addition to all these advantages, the site is now cleared for us by the late fire, and invites to a commencement of this useful, this splendid improvement."[58] The architect was James O'Donnell, best known at that time for additions made to Columbia College in 1818.[59] This structure was designed to be the centerpiece of New York City's public markets, and it went up in only a year. The official opening was on January 22, 1822.[60]

2

THE EARLY DAYS OF FULTON MARKET

When Fulton Market first opened to the public, it was one building, located on South Street, between Fulton Street and Beekman Street. This first building had one story, a basement, and round stone pillars connected by arches. It had rooms over the entrances on either side and the front entrance, where the clerk of the market sat. It had a cupola, housing the bell that the clerk rang to signal the market's closing. The stalls for selling meat, fish, and other foodstuffs extended out from the eaves of this main building.[1] John Pintard, the merchant who was also the founder of the New-York Historical Society, called Fulton Market "an elegant, quadrangular structure, superior in accommodation perhaps to any thing of the kind, probably even in Europe." He also noted the "abundant display of every variety of Meats, Fish & Game, exceeding anything I Have witnessed in this city."[2] The space underneath the building housed mostly saloons that specialized in selling oysters.[3]

Just ten years later, many of the butchers who had occupied most of the original market stalls had vacated their leases because they couldn't pay their rents. Those who remained in the east wing of that building were moved to make room for a special section designated for fish.[4] By 1831, the fish business at Fulton Market had grown large enough that many of them moved to a new building on the East River between what are today Piers 17 and 18. It was a 195-foot long one-story shed, built for $3,123. Some fishmongers remained in the original building.[5] This first shed was demolished in 1844.[6] Both the fish dealers and the produce dealers liked the area because of its excellent access to multiple waterways. The fish came in right off the boat and the produce arrived rapidly by boat from Long Island.[7] The fish dealers survived by selling to other markets around the city. As their neighborhood became less residential over time, the butchers had trouble competing with retail meat dealers who were spreading throughout the city during this era.[8]

In 1847, the city built a new building on the site of the original retail market.[9] In 1853, butchers' stalls still occupied three sides of the main building, but the center included dealers who sold poultry as well as a few who sold dairy products like butter and cheese. Oyster stands now dominated one whole side of the building.[10] The main Fulton Market building had three other brick buildings attached to it. Each held retail establishments, such as restaurants and liquor stores, as well as boarding houses on the upper floors.[11] There was a retail fish market in the main market building, but by the late 1850s wholesale fish dealers had begun to occupy the sheds across South Street. You could buy whatever fish was available at the retail stalls, but the wholesalers specialized in the popular fishes like cod and sea bass.[12]

By this time, the market building was beginning to fall into disrepair and did not always leave the best impression on visitors. The buildings that made up the market were supplemented by a series of tables laid out inside the buildings and in the streets between them. There were also eaves and awnings between the structures that made it difficult to tell where the buildings ended and the outside began. *The New York Times* described the result this way in 1858: "Externally it presents the appearance of an aggravated shanty—a pile of wooden piggeries set up against brick walls that had heard the summons to fall, but could not quite consent to go yet."[13] Complaining about the access going to and from the nearby Brooklyn ferry, *Frank Leslie's Illustrated Weekly* observed, "Pigs escape from the stye, that festers on the river-side of the street and wend their confused way among the legs of enraged men, occasionally upsetting two or three apple-women who instantly add their shrill squalling to

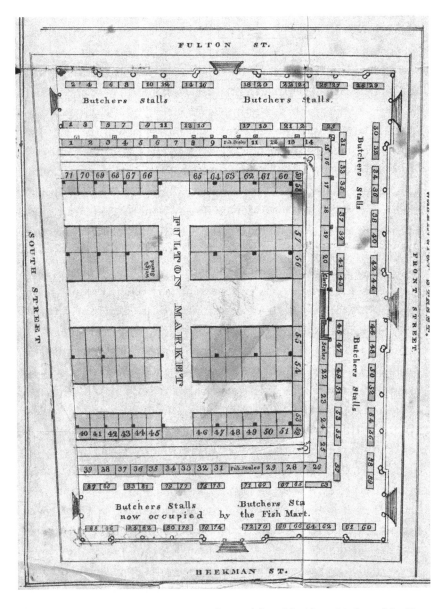

FIGURE 2.1 Map of Fulton Market. From "Groundplan of the Eleven Markets of the City of New York," by Imbert's Lithographic Office, 1827.

Museum of the City of New York

the already deafening sounds of motley confusion."[14] Fish dealers never had to worry about their product escaping even when it arrived at the market alive.

The area around Fulton Market was one of the most active ports in the United States at the time the market was founded. In 1813, Beekman Slip became the Manhattan terminus for the Fulton Ferry, which traveled regularly between Lower Manhattan and Brooklyn. The Black Ball Line, the first regular packet line to Liverpool, England, carried both passengers and freight and docked just below Peck Slip. Later, the Collins Line of transatlantic steam ships was based nearby. This made the neighborhood a central hub for cotton on its way from the American South to Great Britain and for capital (letters of credit, etc.) going in the other direction.[15] "The city side of the East River," wrote an English visitor in 1846, "[is] covered, as far as the eye can reach, with a forest of masts and rigging, as dense and tangled in appearance as a cedar swamp."[16] Starting in the 1840s, the neighborhood surrounding the market began to fill up with businesses related to the ships that docked nearby: sailmakers, figurehead carvers, riggers, etc. Evidence of the trade with China and Japan appeared throughout the area.[17] This trade helped the city itself grow, and its larger population consumed more seafood. While the area prospered commercially, the neighborhood around Fulton Market lost residents. The city itself grew quickly during the mid-nineteenth century, but the population headed elsewhere. Improvements in basic services like water and sewers in this neighborhood lagged behind the rest of the city.[18]

That may help explain why it took decades for Fulton Market to grow into its role as an important retail hub and even longer for it to dominate the sale of fish in the city. With few choices for preservation, customers had to pick up and transport their fish out of the neighborhood, and the longer and farther they had to go, the harder that became. Nevertheless, over the course of the nineteenth century, the fish provisioning system that developed at Fulton Market became very lucrative, working at a large volume. In 1853, about 150 vessels arrived at Fulton Market each day carrying between 20 and 120 tons of live fish.[19] The *New York Herald* estimated that the total receipts at the market that year amounted to $1,126,804.[20] In Fulton Market's early days, people came to do all their shopping—not just for fish. The wholesale market was an offshoot of the retail fish trade. Not until the late nineteenth century did wholesale fish dealers become the most important businesses in the entire facility.

The rules for Fulton Market, passed by the City Council even before the market opened, reflected the prominence of butchers there. For example, the language in those rules repeats the phrase "no butcher or other person" at the beginning of many provisions. Both that language and the sheer amount of space devoted to rules regarding meat cutting demonstrate their prominence at that time. Although there were plans for a separate fish market even then, the butchers were the clear beneficiaries of this first facility. The rules allowed only licensed butchers to cut meat inside the market, which was a way to ward off unlicensed butchers who were spreading throughout other parts of the city, and banned diseased animals and roadkill from being cut up inside the market, which served as another form of quality control. The same rules set up a system requiring the sharing of scales between butchers to prevent anyone with a stall from short weighting their customers. The rules even reiterated already-existing rules against selling meat outside the public market system in New York City and included regulations for area slaughterhouses.[21]

Butchers had economic and social standing. Their stands were the most lucrative spaces in any market because meat cost so much compared to other foods. Another section of the rules read, "It shall be the duty of the Market Committee, and they are hereby authorised, to lease . . . at auction or otherwise, as they may deem expedient for one or more years, all the butcher's stalls or stands and so many of the stands for fishermen, country people, and sellers of vegetables and fruits in the said Market," but the butchers' stands were the only ones actually auctioned.[22] After protesting the first auction in Fulton Market, 88 butchers rented space in the market right before it opened, paying as much as $455 per year for the privilege. Total annual rents on those spaces amounted to $18,865.[23] Because Fulton Market cost so much to build, the city's original plan was to sell the stalls off at auction whenever the contracts ended so that the rents could help cover the city's initial costs.

Starting with Fulton Market, every market in the city began to auction off premium stalls, which were invariably held by butchers, as a way to get at their power.[24]

The economic position of butchers worsened throughout the antebellum era. As early as April 1822, many had deserted their stands, while others began to demand rent reductions. Some were arrested for failure to pay their debts to the city.[25] A yellow fever epidemic in Lower Manhattan during the summer of 1822 did nothing to help business, as all the butchers in the market moved to safer neighborhoods.[26] In 1833, the city's market commissioners created a map of all thirteen public markets in the City of New York. This map wasn't a good way to find them, as it didn't place the markets in relation to one another on the island of Manhattan. Instead, it was a ground plan, designed to display the relative size of each market as well as the number of stalls and the types of food sold by the merchants who occupied them. There were stalls that sold fish in many markets at that time, but only Fulton Market had a section of twenty-seven stalls labeled "Butchers Stalls" with smaller print below that read "Now occupied by the Fish Market."[27] The situation grew even worse for butchers in the 1840s when consumer pressure forced the municipal government to allow meat to be sold outside the public market system. This created revenue problems for licensed butchers by lowering the cost of their product.[28]

As the situation for most butchers worsened, tensions between them and the fishmongers escalated. The butchers attacked the fishmongers as a nuisance based on the traffic they created, the smell of their product, and water leaking into the underground cellars in the original building. In 1842, the butchers complained about the volume of steamboats landing at Fulton Market slip that were fully laden with shad. They blocked other craft with other supplies from provisioning other stalls at the market. Because many dealers sold their fish straight off the boat, the crowds of buyers made moving around the piers exceedingly difficult. Yet in 1843, the butchers begged the fish dealers to return to the main retail market building because the butchers could not attract enough foot traffic on their own. The fish dealers turned them down. The next year, another complaint argued that the growth of the fish portion of Fulton Market was solely responsible for a large loss in business elsewhere in the same facility. In 1844, the fish wholesalers protested an effort to get them removed solely to the "east wing of the Fulton Butcher Market," arguing that it was a "severe inconvenience to them." That move happened despite the protest, but the city did make improvements on that section of the building to soften the blow.[29] While butchery was becoming decentralized around the city, fish wholesaling was concentrating in Fulton Market. This gave the dealers an increasing amount of power to control the operation of the market as time passed.

The butchers remained a constant source of trouble at Fulton Market for as long as any remained there. Even after the city repealed the law that restricted meat sales to public markets and the wealthiest butchers began to open their own shops, Fulton Market retained some, although they never held as much power as they had during the market's very early history.[30] After 1850, butchers turned over frequently. Since the area around the market had ceased to be a residential neighborhood, there were very few people around to buy what they were selling. The butchers who remained did business mainly with the hotel and steamship trade.[31] In 1853, there were 65 butcher stalls at Fulton Market, selling about 16,000 head of cattle.[32] Because slaughterhouses were located elsewhere, the meat cut in the market was never as fresh as the fish, which were kept live in fish cars right outside the buildings. Beef had prestige, but whether pickled, salted, or just old, it was seldom fresh. Fish invariably held a freshness advantage for those who lived close enough to take it home and prepare it quickly. If you bought your meat from one of the less-trustworthy vendors outside of the market, it tasted even worse because even more time had passed since the animal was killed.

As the example of the butchers demonstrates, there was plenty of other economic activity around Fulton Market during its early years besides the sale of fish. In this way, it resembled every other public market in the city. Because both people and a wide variety of trade goods arrived at this area, the neighborhood included many warehouses to store those goods and counting houses charged with keeping track of them. The rest of the neighborhood was populated by a variety of businesses related to the ships that arrived on the pier: chandleries, sailmakers, wood carvers, etc. There were other areas closely tied to different industries. The Flour District was to the south, where the barges carrying flour arrived. Farther south still, close to Wall Street, was where the boats carrying cotton, sugar, and tobacco from the South arrived. Closer to home, Fulton Market was a place for New Yorkers to obtain perishable foods during its early years. While larger vessels docked all along the East River on the Manhattan side, there was a special section near Fulton Market where smaller vessels carrying produce from Long Island or New Jersey unloaded their cargo. Until New York's railroad system grew later in the century, produce usually arrived via wagon or barge and was distributed for sale at any of the city's many public markets.

The opening of the Erie Canal in 1825 vastly improved the range of foods available in New York City, including those in Fulton Market. Historians tend to look at the completion of the canal as an essential step in the opening of what we now know as the Midwest—and it was. But the Erie Canal also built cities like Rochester and Syracuse along its banks, as well as New York itself. Under its influence, northwestern New York State filled in with farmers,[33] who now could vastly extend the range in which they sold their products. While the east–west trade along the canal was the most obvious area for expansion, access to the western market helped concentrate trade of all kinds in the city. Immigration, and the international exchange that accompanied it, made the Erie Canal possible by generating the funds needed to build it. The completion of the canal assured that much of the rest of the East Coast trade had to go through New York City.[34] This success generated the capital that helped build the railroads that dominated trade of all kinds in later decades. Even more than the canal, those railroad lines directed trade through or to New York City.

Visitors to Fulton Market often wrote in detail about the huge variety of products available there. In its early years, these descriptions included more than just fish. This is just a piece of one of the many such lists, from the *Evening Mirror*, published in 1850:

There were Valparaiso pumpkins, manna apples from Cuba; peaches from Delaware; lobsters from the coast of Maine; milk from Goshen; chicken from Bucks county, Pennsylvania; hens from Cochin China; potatoes from Bermuda; peas, beans and squashes from Long Island; whitefish from Lake Michigan; there was beef that had been fattened on the banks of the Ohio; hams smoked in Westphalia; sausages stuffed in Bologna; mutton from Vermont; and cheese from the region of the St. Lawrence.[35]

In 1854, Fulton Market housed more than a quarter of the city's fruit vendors, and nine fruit wholesalers operated from the Fulton Street pier.[36] The foods that arrived from closer to New York City would have been fresh; much of what came from farther away probably wasn't. Either way, the vast variety of food coming through Fulton Market was both a reflection and a cause of the city's rapid growth during this period.

Because transportation technology improved over the course of the nineteenth century, the delivery of foodstuffs became both faster and cheaper over time. After the Erie Canal, new railroads opened up still more farmland to New York City marketplaces starting in 1845.[37] This in turn opened up more land for growing everything, which then fed the expanding population of cities like New York. Preservation methods also improved, which increased supply and decreased price. Fulton Market started life as a general market for foods of all kinds but came to specialize in fish because it was more efficient to concentrate fish dealers in a single place. Infrastructural improvements, including the new building right next to the water, made it possible for fishermen to come right up to the market to unload. This gave a huge boost to the wholesale fish industry in the 1850s.[38] The wholesalers could coordinate the nature of both supply and demand, allowing both the fishermen who caught fish and the restaurants that sold them to maximize their profits.

Many vendors during the market's early years were country people from surrounding farms selling products they had grown themselves. They rented tables at the market by the day on a first-come, first-served basis. They had to scramble for space every morning because there were always too few tables to accommodate every vendor.[39] Initially, large-scale butchers rented the many cellars in the building for storage, but those spaces increasingly went to other, poorer vendors who sold foods of all kinds as the number of butchers in the market fell.[40] During the fruit and vegetable seasons, when these vendors crowded their place each morning, it would have been difficult to pass.[41] Some of the successful tables became stalls even though the city had once tried to drive all of them away.[42]

The increasing number of restaurants throughout New York City took advantage of all that newly available food. Around 1830, the city had grown large enough that the separation between home and work required people to go out for their midday meal. This had once been the largest meal of the day, but workers of all kinds came to prefer a quick snack, known as "lunch." One of the quickest and most popular snack lunches was oysters, like the ones sold in the stands at Fulton Market.[43] Many of the oyster stands grew into full-blown restaurants. By 1883, restaurants took up about a quarter of the entire space of the market's retail building.[44]

So-called "coffee-and-cake shops" offered public dining options for the poor and working class. Most of these places had a bad reputation.[45] However, the ones at Fulton Market were well known for serving a grateful dinner crowd with pies. "Pies are a great institution in Fulton Market," reported *Macmillan's Magazine* in 1867. "The number annually consumed at the various coffee-shops here is known only to the fortunate manufacturers." Many of them were sweet, but many, designed to be meals by themselves, were savory. "They are sometimes served hot, but their normal condition is one of cold leaden clamminess. About the middle of the day, *cartloads* of them arrive at the market, and they are distributed among the various coffee-shops. At the usual working-man's dinner-hour, these places are filled to overflowing by hungry pie-seekers, who come to enjoy their dinner."[46] The coffee-and-cake shops in Fulton Market catered to late-night passengers to and from Brooklyn on the nearby ferry.[47] The most successful stalls took on the trappings of restaurants over time.

The totality of the retail possibilities at Fulton Market was hard to summarize. "It is vast on candies," wrote the *New York Herald* during the Christmas season of 1868, "immense on peanuts, extravagant on coffee

FIGURE 2.2 Game stands at Fulton Market. For much of the nineteenth century, Fulton Market vendors sold a lot more than just fish. "The Christmas season / game stand, Fulton Market," by A. B. Frost, *Harper's Weekly*, January 5, 1878.

The Miriam and Ira D. Wallach Division of Art, Prints and Photographs: Print Collection, The New York Public Library

and cakes, captivating on oysters, prime on fish, desirable on lemons, the buyers always getting squeezed; luscious on oranges and enormous on tenderloins and sirloins, and, like the aged establishment it is, retires early. . . . At midnight, that is."[48] The *New York Times* wrote colorfully about the variety of retail experiences there in 1879. "Recently, the market has not been as bad it was," the paper concluded. "Still, it is a nondescript structure with its groggeries, coffee and cake shops, petty haberdasheries, cheap restaurants, hucksters' stalls, oyster-saloons, news-vendors' stands, and skirmish lines of fruit-hawkers wagons. It abounds in placards of various descriptions such as a very hungry and unfastidious man, with a few nickels or coppers in his pocket, finds difficult to resist."[49] Most New Yorkers never visited the wholesale fish building, which became increasingly important to the economic health of the market as it transformed into the place for buying and selling fish.

The roots of the wholesale fish emporium that Fulton Market became were clearly visible during this era, but nothing about the modest operations in the facility's first decades made it look like an all-fish future was somehow inevitable.

Although Fulton Market became a tourist destination early in its history because of the vast variety of food available there, by the mid-nineteenth century it remained a tourist destination mostly because of its close association with fish. In 1856, Theodore Gill, an ichthyologist then working for the Smithsonian Institution, reported on the regular visits he had made to Fulton Market to document the range of species he saw there. His intention was to record at what time of year each species first appeared for sale and when it disappeared, in order to learn more about its life cycles and migration habits. This effort proved only partially successful because of the huge variety of fish. "You will notice in my catalog that I mention seventy-nine species in fifty-six genera and twenty families," he reported, "all of which I have myself seen." He went on to confess, "The great variety of species has surprised me." That variety outnumbered the known species ever seen in New York's local waters.[50]

This was a byproduct of the gradual transition from a predominantly retail market to a predominantly wholesale market. Just collecting fishmongers in one place increased the efficiency of distribution, even if the fish were handled inefficiently once they arrived, because fishermen knew they could sell their catch there, whatever it happened to be on any particular day. It helped that the market was along the water, so that the fish could travel farther while still alive and be stored longer in the many fish cars. For similar reasons, it was more efficient to process whatever fish New Yorkers would eat in that single place. As more New Yorkers began to shop in grocery stores, which were distributed far more conveniently in neighborhoods across the city, Fulton Market became a new link in the provision chain by specializing in many different types of fish.[51] The people in the neighborhood and those who arrived daily on the ferry from Brooklyn were becoming less important to its overall operations.

The easiest way to find out what the fishmongering experience was like in the early days of the market is to read the biographies of people who succeeded in the business, written when they looked back upon their careers later in the nineteenth century. Many of the earliest fish dealers in Fulton Market created companies that persisted into the twentieth century in some form. Many of the oldest firms were founded by fishermen who came down from Stonington, Connecticut, where their boats had once been based. Sam and Amos Chesebro (originally Cheeseborough), who formed Chesebro Brothers, Roberts and Graham in 1893, followed this path. The firm grew to be the largest in the market during the 1950s.[52] Plenty of others failed, but the ones that survived over decades came to define the culture of the fish market, as control over them often stayed within the same family.

Many earlier Fulton Market firms were organized by people whose jobs involved the retail sale of fish rather than fishing itself. Dudley Haley, for example, started his wholesale fish business in 1845. In 1852, he was successful enough to be elected alderman. He had very little experience as a fisherman and gained his knowledge of fish simply by buying and selling them for a long period of time. In 1865 he brought his son Albert into the business, which was then called D. Haley & Company.[53] That company was succeeded by a firm called Burnet and Keeney, which in 1917 was sold to the Atlantic Coast Fisheries Company. A related firm lived on under the moniker Burnet and Keeney Continental Company.[54] Another firm, Packer and Haley, was started by Caleb Haley, who had learned the business by keeping books for his uncle Dudley. In 1862, he married Elizabeth Miller, the daughter of another Fulton Market pioneer.[55]

Samuel B. Miller started working at the Fulton Fish Market in 1833. His father, a weaver, had decided to try his hand at the fish trade in 1827 and brought his son into the business when he turned thirteen. The younger Miller founded S. B. Miller & Company in 1841; he added his brother Charles to the firm in 1851 and later his two sons. Like both Haleys, Miller had no experience on fishing boats.[56] He devoted much of his career to getting the public to appreciate different kinds of fish.[57] In 1895, when Miller had been working at the market for sixty-two years, the *New York Sun* reported, "Catch him at a time when he isn't busy and he'll spin yarns, tell of the old days, give the history of all the men and boys in the market from the time they cleaned their first fish, and when he's asked about fish, he can talk for hours."[58] These were small firms with few employees, many of whom were children. As the similarity in names suggests, many came from the same families. That promoted a culture of ethnic, racial, and class solidarity in the market that helped insulate the industry from the outside world. Before long, the similarity among the men who operated all these firms led to formal organization.

In 1869, the largest firms organized the Fulton Market Fishmongers Association in order to pool their money and lease land from the city on the waterfront on South Street between Beekman and Fulton Streets. On that land, they constructed a new building specifically designed for their wholesale operations, which opened the same year. It was a two-story wooden building topped by a cupola. The first floor had seventeen new stalls for fish sales. The second floor had offices for the association, storage space, and a reading library.[59] The previous fish-related buildings

FIGURE 2.3 New Fulton Fish Market Building, 1869, by Theo. R. Davis. This was the first permanent structure in Fulton Market devoted entirely to selling fish.

Library of Congress, Washington, D.C.

had been just wooden sheds; this was intended to be a permanent structure. The new building was a sign that the local wholesale seafood industry had come of age.

The new organization that operated the building was a sign that the fishmongers had enormous control over the buying and selling of fish of all kinds. As long as they could stick together, the wholesale dealers held monopoly power over the fishermen who sold to them and over New Yorkers who wanted to buy most kinds of seafood. "Having grown from an humble origin to its present dimensions," wrote the *Times* in 1870, "the worthy fraternity [of fishmongers] began to feel its power, and determined to make a more respectable show in the world."[60] The permanent structure offered respectability. So did the decision of the State of New York to charter their organization. The bylaws of that charter allowed a majority of the fishmongers to impose their will upon a minority if disputes arose between them on many issues. This was a way to bring unity to an industry of many comparatively small actors with multiple competing interests.[61] Neither the fishermen who sold to the dealers nor the retailers or consumers who bought from the dealers had any such vehicle for unity among themselves. While the city was still involved, the high degree of regulation in public markets like this one gradually fell.

Fishermen, knowing that they had a reliable market for their catch, began to overfish the waters near the city. Consequently, they had to travel farther away for sufficient amounts of product to sell. Although there had been overfishing even before Fulton Market opened, the problem grew worse over the course of the nineteenth century as increased demand met a supply of fish limited for other reasons. The first victims of these combined problems were freshwater fish, whose populations were already stressed because of changes to the river systems that they depended upon. In response, the fishmongers in Fulton Market did their best to manipulate demand for specific species so that the profits would continue. By doing so, they permanently changed what Americans ate.

3

FISH FROM FAR AWAY

Compared to the bounty taken from the waters around New York City in colonial times, the situation in the early nineteenth century seemed dire. As late as 1832, sea bass was the most popular fish at Fulton Market, and most fishermen did not have to travel past the southern part of New York Harbor to find them.[1] However, just six years later, Peter Cortelyou, a member of one of the early Dutch families that controlled the local fisheries, complained, "All the fisheries in New York harbor are nearly destroyed, and the fish that now supply the markets of that city are brought from the distance of sixty, eighty, and even a hundred miles."[2] In the same way that historians suggest that the Erie Canal gave New York City a kind of imperial chokehold over much of the grain produced in the Midwest, the growing importance of Fulton Market to the trade in fish and seafood of all kinds gave the city a kind of imperial control over water—oceans, rivers, lakes, and streams. Those distances could be a problem for boats that had no way to keep their catch fresh. The use of ice to refrigerate fish made this possible.

That imperial control only improved as refrigeration and freezing technology got better. "The inhabitants of New York . . . have the greatest command of fine fish that is possible to any one city," reported the *Times* in 1872:

> Throughout the year the [fishing boats] of Gloucester, Marblehead, Nantucket, Martha's Vineyard, Noank, of the whole coastline and islands, in short, of Massachusetts, come up into the East River to the Fulton Market to dispose of their merchandise. Not alone these, for the more adjacent fishing towns of Long Island or New Jersey, bring their freights, and even the schooners of Canada and of Newfoundland have for some time past begun to turn their bowsprits in this direction.[3]

A little later, red snappers from the Gulf of Mexico began to appear regularly in the market, as well as salmon from the Pacific Northwest.[4] Better refrigeration and effective freezing changed how consumers perceived seafood of all kinds and greatly increased the consumption of fish and shellfish throughout the city. That's why freshness actually improved as fish arrived from farther away, as did the variety of choices for consumers. The greatest financial beneficiaries of these changes were the wholesale fish dealers in Fulton Market.

Ice became the key to transporting fish over long distances as it became increasingly accessible over the course of the nineteenth century. It became more popular faster on the part of the provisioning chain over land rather than the part over water because fishing boats tended to keep their catch alive if they operated in smaller volumes. Nonetheless, fish transported on ice remained fresh no matter where they happened to have died. Ice keeps fish better than any other form of preservation except for flash freezing, because the water from melting ice keeps the flesh more moist.[5] Cold-water species spoil faster than warmwater species both because of the fatty acids in the fish from cold water and because the bacteria that break down a fish thrive in cold temperatures. Mackerel and herring placed on ice will remain in good condition for only five days. Cod and salmon will remain that way for eight days. Trout can last fifteen days.[6] These were the limits of the fish supply chain before cold storage and flash freezing. The distance from port and the speed of the vessel the fish were landed in determined one half of the chain's length, while the distance and time to the point of consumption determined the other.

The ice industry began in 1806 when a Boston merchant named Frederic Tudor cut some off of Fresh Pond, near Cambridge, Massachusetts, packed it on a ship, and tried to sell it on the Caribbean island of Martinique. His subsequent efforts were marked by hit-or-miss experimentation with cutting instruments, ice houses, and various ways for customers to keep ice in their households without it melting too quickly.[7] Improvements in ice cutters and ice houses in widespread use by 1827 provided more ice for other uses like transporting fish. Ice produced for use in the city came from the Hudson River and places like Rockland Lake in Rockland County. By 1856, the total output from firms that sold their product in the city was 285,000 tons.[8] The supply grew as the century progressed.

It took time for Fulton Market's sellers and buyers to fully take advantage of the availability of ice. There had been no market for dead fish in New York City before 1825.[9] Fishermen took small loads from local water. Ice expanded that range. "Last week, about five hundred weight of fresh salmon, from Lake Ontario, was sold at Fulton Market at twenty-five and thirty cents per pound," reported the *New York Mirror* in 1826. "They were conveyed to this city via the Erie canal, packed in ice, and in fine order."[10] Shipments like that got notice in the newspaper because they were still quite rare. The first ice was used on fishing boats around 1840. At first, the ice was kept in rolling rooms on board separate from the fish, but by 1846 fishermen started packing the fish in crushed ice because they discovered the preservative properties it had for the skin.[11] This meant that fish caught farther away appeared in the market for the first time.

The growth of the railroads made it possible for fresh fish kept on ice to appear more regularly. With enough ice to supply both fishing boats and railroad cars, fishermen didn't have to land at Fulton Market in order to sell there. Instead, they could land elsewhere, have the fish processed at that landing place, and then have it shipped to Fulton Market in ice for sale in and around New York City. How to ice a fish for shipment seemed obvious, but wasn't. How much ice to use depended upon the time of year, the distance the fish was going, the kind of fish, and the quantity. It also helped if the cargo was prechilled, rather than just packed in ice after sitting at air temperature for any length of time.[12]

Ice was also important for sending seafood out of Fulton Market. Many fish were landed at the pier, unpacked from the ice compartments of boats, and packed in ice boxes or barrels. "Along one side of the building, a swarm of men work like beavers," the *New York Times* reported in 1884, "paying no regard to the buying and selling. They pack the fish in long boxes with powdered ice and start them off on huge drays for the railway station." The total time in the market could be as little as two hours; then the boxes left New York for customers in Baltimore, Boston, or even Nevada.[13] By icing that fish in barrels or boxes after it arrived, dealers in Fulton Market could use railroads to sell their fish over a larger range. They could also ship that iced fish via parcel. For the first time, people could eat fish that had never been available fresh near where they lived, far inland. Because the United States was so far advanced in the production of ice compared to other countries, people who lived inland could eat fresh fish from the ocean much sooner than people who lived far away from the sea in other parts of the world.

Because ice was expensive or impractical for much of the nineteenth century, a lot of the fish leaving Fulton Market during its early years was dried or salted. With very few restaurants in the city, most customers took home dried or salted fish.[14] This situation lasted longer for places outside of major cities. A popular cookbook from 1873 noted that "country housekeepers . . . can seldom procure fresh cod," perhaps the most popular fish in the country at that time. Covering it with heavy sauces or mixing it into fish balls were ways to prepare fish preserved by methods that changed its taste or texture.[15] Salt cod had to be soaked for days to get the taste of the preservative out of it. Fresh fish, when obtainable, often had to be cleaned by the cook in the kitchen because processing did not yet occur at places like Fulton Market.[16] New York City, where the chain was much shorter for any ocean fish, had fresher fish that usually had less time to decay on the way to people's plates.

The increased use of ice decreased waste. That increased supply, which in turn made fish more affordable. No wonder wholesale fish dealers needed greater reach in obtaining their supply. "The fishing fleet is continually searching new grounds or exhausting old ones," wrote a reporter for the *Times* in 1888.[17] This made up for the loss of fish closer to shore even as the fishing fleet depleted new stocks located farther away. Places like Fulton Market reflected the impact of this fundamental change. As fish populations dropped in particular locations, the supply in the market would often disappear, to be replaced with a more available species. Oftentimes wholesalers would start to buy similar species from different places whenever those fish became available.

People who lived near the ocean could eat fresh fish because it came to port alive in a fish well. Fish wells, as Thomas De Voe explained, were "placed in the centre of the vessel, by which sea-water can flow in and out through a latticed bottom, and thus preserve the fish alive and fresh."[18] There was also a "well deck," which limited the height that the water could reach in order to keep the boat steady if it was rolling.[19] Fish wells originated with the Dutch fishing fleet, but the English copied this innovation during the seventeenth century. The first boats with wells appeared in New England around 1820.[20] The problem was that they only worked during the cold months. Fish caught in colder waters off New England and carried in fish wells into warmer waters near New York would die.[21] Luckily, fish wells could also hold ice. The name for any boat with a fish well was a "smack." Although this is a generic term for a small fishing boat, during the nineteenth century the presence of a fish well was the defining trait of a fishing smack.[22] Almost every fish dealer in the market owned or at least had an interest in a fishing smack by the mid-1870s.[23] The wholesalers became vertically integrated to assure a steady supply of fish for their buyers and to make it easier to keep their prices down.

Getting fish from farther away required larger ships to make longer trips and have the space to return with the catch. The smallest vessels serving Fulton Market during this era were oyster sloops, about 35 feet in length, mostly working the waters on either side of Long Island. Larger

FIGURE 3.1 "The Fish Landing Strip at Fulton Market, New York," by D. C. Beard. Fulton Market owed its waterfront location to the fact that fish could be taken off boats alive and stored or processed there with little trouble. *Harper's Weekly*, April 30, 1887.

Author's collection

sloops of around 35 feet in length operating out of Connecticut or New Jersey caught cod or mackerel farther out at sea and returned to Fulton Market to sell it. Market schooners of 100 feet or more went even farther, sailing out to Georges Bank off of Cape Cod, where they utilized newer fishing methods to get even larger catches of fish like cod, bluefish, or sea bass. In the right season, they would travel off the coast of North Carolina and follow schools of fish as they migrated northward.[24] By 1870, more than 250 such vessels appeared in Fulton Market regularly. "They visit all banks shoals and streams within a radius of a thousand miles," reported the *New York Times* that year, "and gather the tender victims by the ton . . . Salmon from the Kennebec, shad from the Connecticut and trout from many an island stream are brought per force from their cool retreats, while scores of white-winged smacks are probing the shoals of Nantucket and the banks of Newfoundland for the more bulky cod and halibut."[25] Fish caught at these farthest destinations would have been landed at a port like Gloucester or Boston, then killed and packed on ice and shipped to Fulton Market by train.

Fish wells were a very effective means of preservation because live fish don't decay. Pick a watermelon, even if it's not ripe, and the countdown to the moment it becomes inedible starts immediately. The result, when shipping produce from far away without refrigeration, is a lot of waste.[26] The same thing happens when you kill a fish. No wonder fishermen resisted turning to ice for so long. Ice slowed the process of decay but could not stop it. That meant wholesalers had to have a provisioning chain speedy enough to get their catch to market long before it spoiled, so that there would still be some time on the other end for the fish to be cooked and eaten. What changed storage techniques was improvements in fishing methods that allowed fishermen to increase their volume. The more fish a smack could hold, the farther away it could go to catch them. But eventually fishing techniques outpaced the size of the fish wells on the smacks that had once depended upon them.

Starting in the mid-1850s, the fishing smacks began to fill their fish wells using a method known as trawling. Irish immigrants to the Boston area introduced the practice of catching fish on what was known as a trawl line to the American marketplace and proved very successful at it.[27] A trawl line was a very long fishing line, with short branch lines known as snoods every three to six feet. A trawl line was between 3,600 and 7,200 feet long. The snoods were usually three feet long. Each had a hook at the end of it, making it possible for a small number of men to keep a lot of hooks in the water at the same time. The smaller lines rested on the surface while the long line was kept afloat by buoys.[28]

The fishermen would do this kind of fishing from dories, small boats stored on the main vessels. They would first throw bait directly into the water in order to attract a school, then cast their lines. At that point, the men in the dory just keep reeling in fish until their smaller vessel filled up or the bait ran out.[29] Each dory contained a single (or later, a pair) of fishermen, a long trawl line, and plenty of bait. They would stay out for hours at a time. On a good night, one man could hook three or four hundred fish at one go. Using dories to catch bluefish or cod allowed a single smack to cover an enormous area of the sea. Including the length of their lines, the men from one fishing boat might reach fish over an area of six to eight square miles.[30] Later in the century, when ice became a necessity, a smack might carry fifteen or twenty tons of it when departing port. Bluefish were usually caught off Cape Hatteras in North Carolina, a very long trip.[31]

Hand line fishing with dories began off the coast of New England during the late 1850s. Every man on the boat had a dory of his own to fish from except the captain and the cook. Each man tended two lines from their smaller vessel. There might be six men and a cook on the typical smack. The trip east to Georges Bank or the Grand Banks of Newfoundland—both well-known fishing grounds and far out to sea—took about six days. If the boat was having good luck, the men might get four or five dory loads each day, and the boat might be anchored in a single spot for several weeks. In the right season, there might have been 500 or 600 dories within sight of one another, all working the same gigantic schools of fish.[32]

Cod was the most popular fish for mid-nineteenth-century smacks to hunt. Long before ice came along, fishermen caught cod in large numbers, then dried or salted them so that they would hold up as an article of commerce. Atlantic cod are all generally the same, but sometimes get divided into species like "Icelandic" or "Arcto-Norwegian" that exhibit minor variations.[33] Cod, wrote De Voe, "is quite extensively known, and always to be found in our markets. When fresh, its flesh is white, firm, flaky and very good."[34] In the mid-nineteenth century, cod was on every stand and made up about 25 percent of all the fish sold at the market. In 1853, it could still be found off the coast of New York and New Jersey during the winter; those fish were landed in Fulton Market alive. The codfish from farther away was either salted or packed in barrels with ice.[35] Later in the nineteenth century, after trawling began in earnest, fishermen working out of Gloucester or Boston could land cod in Massachusetts, pack them in ice, and send them off to New York by rail nightly in order to meet demand.

Cod fishermen started to experiment with dories and long lines during the mid-1850s. As the volume of their catches increased, even New England fishermen had to travel farther because the fish stocks closer to their shores became severely depleted. Farther out to sea, they had to compete with French trawlers covering the same areas. The cod catch crashed not too long after.[36] Cod migrate long distances depending upon the season, the water temperature, and other factors that weren't fully understood during the late nineteenth century. However, it was clear as soon as the early 1870s that fewer were to be found near the coasts after new fishing methods were employed to catch them.[37]

In this era, fishing smacks were generally owned by groups of three or four people (often including wholesale dealers like the ones at Fulton Market). This was an important way for dealers to learn where and how to get fish. Usually the captain of the boat had a stake in its success. If a crewman owned a share in the boat, he was entitled to two different shares of the profit: one for his capital and the other for his labor. Very few of these boats were based in New York, but there were 130 such vessels owned fully or partially by Fulton Market dealers during the mid-1870s.[38] Many of the successful dealers at Fulton Market got their start on smacks based elsewhere and moved to New York City after they had earned enough money to begin buying and selling wholesale. In 1872, a consortium of Fulton Market dealers set up shop in Gloucester as the New England Company and sent much of their catch down to New York themselves.[39] As long as the provisioning systems ran through New York at some point, Fulton Market remained at the center of the wholesale fish trade in the United States. However, as the technology of both refrigeration and transportation improved, that situation changed.

The technology of catching fish also changed operations at Fulton Market. One of the first signs of a declining fishery is the movement of fish away from the shoreline.[40] With ocean fishes, as supplies dwindled, fishermen had to travel farther and farther to find any to catch. "Now that the insatiable appetite of the human race has thinned out the schools of halibut in the North Atlantic and is driving what remain to the deeper seas beyond the continental slopes, market men are looking to other waters for their supplies," explained an 1893 article. "Halibut from the Pacific Ocean, selling in Fulton Market, are not uncommon these days."[41] By 1913, the fleet serving the Fulton Fish Market had to go as far south as the waters off Florida to find bluefish.[42] Market men were not worried about local disappearances as long as there were other sources of popular fish for their stalls.

With species that swam up rivers to spawn, Fulton Market dealers sourced their fish from rivers that were farther away because they could still arrive in New York City fresh, packed in ice. Salmon, one of the most popular fish in Fulton Market during its first decades, offers a good barometer of how changes in supply affected the provisioning chain that was centered there. The largest rivers of the Northeast—including the Connecticut, the Kennebec, and the Penobscot—were once filled with salmon. "Salmon, Shad and Alewives were formerly abundant here," explained Henry David Thoreau of the Concord River in 1839, "and taken in weirs by the Indians, who taught this method to the whites . . . until the dam, and afterward the canal at Billerica, and the factories at Lowell, put an end to their migrations hitherward. . . . It is said to account for the destruction of the fishery."[43] In that same year, a prominent ichthyologist declared that the entire salmon population in the Commonwealth of Massachusetts had been eliminated.[44] Salmon had also been readily available to fishermen along New England's shores near the rivers they swam up earlier in the century, but the fish disappeared from there too.[45]

Salmon had been common in New York rivers, like the Salmon River in Oswego County, the catch from which appeared regularly in Fulton Market at the beginning of the season until, like the Concord River, it was dammed.[46] As the salmon population faced different obstacles—overfishing being only one of them—the source of Fulton Market's salmon moved farther away. The first Kennebec salmon, from Maine, began to appear regularly in Fulton Market in 1832.[47] Much of the salmon sold in New York by that time originated in Pacific Northwest and arrived in the city after a week-long train trip.[48] Ice could keep it fresh enough that restaurant goers couldn't tell the difference between that and the East Coast article.

Other salmon sold at Fulton Market originated in Canada. In 1876, one of the dealers erected vast storehouses along the banks of the Restigouche in Campbellton, a town in northern New Brunswick, Canada. The railroad between there and Montreal had just been finished, so he bought up the salmon caught there and shipped it back to New York, using ice to keep it between the river and the storage area and snow to pack the fish on its way to the market. This high-quality salmon arrived in Fulton Market more quickly than ever before because the trip took just forty-eight hours.[49] These laborious and complicated efforts were a testament to just how much New Yorkers were willing to pay for good fresh salmon. Canned and smoked salmon from the Pacific Northwest or Canada were both good alternatives when the fresh article proved too costly for consumers.[50]

Shad fishing was indirectly related to the fisheries of these larger fish because the larger fish fed on young shad at the mouths of rivers across the eastern United States that emptied into the Atlantic. Shad, wrote Thomas de Voe, were a "well-known fish" and "a general favorite among all classes of persons as its flesh is considered among the best, sweetest [and] the most delicate."[51] During the late nineteenth century, shad appeared in most North American river systems that emptied into the Atlantic. Shad, like salmon, are ocean fish that travel back upstream in fresh water to spawn. The season on the Hudson began in late March or early April. When shad fishermen depleted the supplies of young shad by catching larger ones before they could spawn, it hurt the supply of the larger fish that ate them too.[52] Shad fishermen left small, twenty-foot-square nets in those rivers, marked by long wooden poles so that boats wouldn't get tangled in them.[53]

The main drawback to this fish was their numerous bones. "Someone once said that if all the bones of a six pound shad were placed end to end in a continuous straight line they would go twice around the world," joked an 1884 guide to the fishes of the Atlantic coast, "and if they were piled up in a heap, it would form a pyramid two feet taller than the pyramid of Cheops."[54] However, if shad is cooked long and slowly,

FIGURE 3.2 "Shad fishing on the Hudson River, raising the net between poles," c. 1899. Shad fishing was more like fish trapping than conventional fishing. Fishermen had to periodically check the nets they set to see if they had caught anything.

Instructional Lantern Slides Collection, New York State Archives, Albany, NY

the bones soften, making the fish easier to eat.[55] Aside from the bones, the tendency of shad to die instantly after being taken from nets made them harder to preserve, especially in the era before ice.[56] That's why, before the opening of Fulton Market, shad were taken in such great quantities that they were often used as fertilizer. Yet the demand never ebbed during the nineteenth century. Shad was always a very popular and relatively cheap fish at Fulton Market. Many people liked it baked, stuffed with bread crumbs, salt, pepper, and parsley.[57]

Shad were so popular that the supply crashed long before the advent of industrial fishing. As De Voe explained in 1867, "This shad-fishery has been gradually decreasing since the year 1824, so that now it is scarcely worth attending to."[58] As a river fish, the shad had a number of different problems—some of which all fish faced and others that were unique to shad. Dams, especially in New England, continued to affect all river fishes and explain why shad never managed to make a comeback in that area. Over time, the locations on the rivers where shad were caught changed too. When fishermen began to catch shad near the mouths of rivers rather than upstream, fewer of them ever made it to their traditional spawning grounds, which inevitably hurt the overall supply. Rivers mattered because fishermen could catch much more fish when they returned to fresh water to spawn than out in the ocean. A general problem, pollution adversely affected shad numbers throughout the entire eastern United States.[59] Writing about the Hudson River in 1869, *Scientific American* predicted, "The shad, if not looked after, will in less than twenty years be a 'thing of the past.'"[60]

The solution to the problem of depleted shad stocks for Fulton Market dealers was to source their shad from farther away. As early as the 1840s, scientists recognized that shad appeared in different rivers at different times. Unlike most of the fish sold at Fulton Market, they spent most of their lives south of New York City and only returned to fresh water to spawn. Spawning shad first arrived near Charleston, South Carolina, in January, in New York in April, and in Massachusetts in May.[61] When shad became rare in the Hudson (or when the Hudson was just having a bad year), dealers would get their supply from farther south. Shad from both North Carolina and Florida that appeared in Fulton Market in 1886 "created a sensation" because of their size.[62] The successful introduction of shad in California in 1871 (it eventually spread to the Pacific Northwest) demonstrates that shad were never threatened as a species, but the demand around New York did threaten their local population.[63]

The local shad were prized on the New York market because they were widely perceived to be the best available, probably because they arrived at New York tables in the freshest state. The first shad of the season caught in the lower Hudson (also called the North River at that time) on its way upriver to spawn was announced in every newspaper. A group of citizens then presented that shad, decorated with ribbons, to a prominent dealer in Fulton Market in a ceremony attended by hundreds of people.[64] All the fuss demonstrates the importance in Fulton Market of fish that were physically closest. As more and more fish arrived from farther away, the dealers still wanted to stay in touch with local conditions, like they had early in the nineteenth century. Local fish tasted better, but everyone's livelihood in Fulton Market increasingly depended upon fish that were caught in faraway waters, using industrial methods that their grandparents (assuming they worked in the market, as many did) would have found unrecognizable.

The contrast between the fish and beef supply chains is once again informative. Cows were a non-native species that were fed in Texas, shipped to Chicago for slaughter, and consumed on the East Coast. The need for a lengthy supply chain came from the distance between the concentration of the supply and the majority of the eaters. The first fish that were consumed in New York swam near the city. Only as local fish disappeared did a longer supply chain become necessary. Similar to cattle, only as refrigeration technology improved did distant markets emerge. By the end of the nineteenth century, fish arrived in the city by other means—first by railroad and eventually by truck. At that point, as eventually happened with beef, there were other places to process the product as the industry decentralized. Much of the fish landed in the late nineteenth century went straight to Fulton Market. Around the turn of the twentieth century, New England seaports like Gloucester began to become hubs themselves, establishing connections with inland cities that bought fish products processed there on shore.[65] Despite increased competition, the wholesale fish business came to define Fulton Market after the turn of the twentieth century. The same thing was true for the oyster business, which became more and more important as the nineteenth century passed.

4

THE HEYDAY OF NEW YORK'S OYSTER INDUSTRY

"The oysters of the market have been celebrated through all this generation as the best that the globe affords," observed the *New York Times* in 1879. "The Manhattanese reason thus: Of all the oysters of the world, those of America take the lead. Of all the American oysters, those of New York stand first, and Fulton Market oysters excel any oysters brought to this port. These oysters have a metropolitan, national, and even transatlantic reputation. Epicures in shell-fish, or in sea-food generally, have long made it a custom, and ranked it as a rare titilation [sic] to the palate to go to Fulton Market."[1] However, during most of the nineteenth century, the firms that provided New York City with its world-famous oysters were not based in Fulton Market. The wholesale oyster trade, including pickling and other forms of processing, was concentrated on the exact opposite side of Manhattan.[2] Only later, as oysters' popularity waned, did the center of the trade move to the Fulton Fish Market.

Oysters are a bivalve, meaning that they live in a hinged shell. Except during their larval stage, oysters spend their lives cemented to underwater surfaces. The fleshy part inside the shells is often eaten alive and prized for its complex flavor profile. While they can be cooked in many different ways, oysters were (and are) one of the few kinds of foods that human beings often eat while the creatures they come from are still alive.

There was an old saying that you should only eat oysters during months with an "r" in them—in other words, not at the height of summer—in order to avoid oysters that were full of water or could cause food poisoning, since the water was more likely to have germs in it at those times of year. This was both a common tradition and sometimes the law in New York State.[3] The real reason for the law was to make sure that there was always seed stock for future oyster growth. The idea that eating oysters during the summer months was somehow a risk to health was a myth. Most New Yorkers in the late nineteenth century ate oysters all year round because the supplies of beef and pork didn't meet the demand for protein.[4]

There are only three important species of oysters in the United States. Two are found exclusively in West Coast waters. The American (or Eastern) oyster is found everywhere along the East Coast, from Prince Edward Island down to Texas. Nonetheless, some of its characteristics, like the growth rate, depend upon environmental conditions,[5] as does the taste. This mostly stationary creature, more than any fish, best reflects the environment in which it grows. Oysters taste like the water that they filter, and for much of the nineteenth century the bivalves New Yorkers ate grew in the waters surrounding their city. In other words, they were the very definition of local seafood.

Because the taste of an oyster is so influenced by the water it filters, the names of different brands of oysters often suggested their points of origin. During the heyday of the New York oyster trade, it seemed as though every bivalve had its own name. Even local oysters were divided into various brands that sold for different prices based upon their perceived quality. The best known oyster brand was Bluepoint, from Long Island, originally from the waters off the town of Blue Point, near the town of Patchogue. Other popular oyster brands tied to local areas of Long Island were East River oysters and Saddle Rock oysters, which came from the water near a rocky outcropping near Great Neck.[6] The water was clear enough during much of the nineteenth century that oystermen harvested in all the waters adjacent to New York City. Chesapeake Bay oysters were also very popular in New York markets. Oystermen imported tiny seed oysters from there to refurbish local beds that had become exhausted, but Chesapeake oysters were also a quality imported product in their own right.

Speaking of the brands available from just one wholesaler, George M. Still, Joseph Mitchell's Old Mr. Flood, remembered, "From New Jersey he had Shrewsburys and Maurice River Coves. From Rhode Island he had Narragansetts and Wickfords. From Massachusetts he had Cotuits and Buzzards Bays and Cape Cods. From Virginia—they were very fine–he had Chincoteagues and Lynnhavens and Pokomokes and Mobjacks and Horn Harbors and York Rivers and Hampton Bars and Rappahonnocks. From Maryland he had Goose Creeks. From Delaware he had Bombay Hooks." With respect to local New York oysters, Still had the brands mentioned above, along with Mattitucks, Robbins Islands, Diamond Points, Fire Places, Montauks, Hog Necks, Millponds, Fire Island Salts, and Shinnecocks. "I love those good old oyster names," Flood explained. "When I feel my age weighing me down I recite them to myself and feel better."[7] The most expensive oysters were the ones that came from their natural beds, but the industry could never meet the demand without seeding beds in other waters around the city.[8]

As this huge variety suggests, New York City was at the center of the country's oyster trade during the late nineteenth century. Early on, processors pickled their shelled oysters so that they could reach distant places intact, eventually including overseas. Later on, oysters in their shells were packed in barrels or boxes with ice. In 1853, the wholesale and retail oyster trades were worth five million dollars a year and employed 50,000 people. The oyster beds themselves were worth twelve million dollars.[9] In 1859, New York City residents spent more on oysters than they did on meat.[10] By 1872, New York's oyster interests controlled a third of the oyster trade in the entire country.[11] By 1892, the New York oyster trade had doubled to about ten million dollars per year. Twenty-nine oyster firms split the local farming and wholesale business.[12]

At this industry's height, about 20,000 New Yorkers owed their employment to oysters in some way. These included people who worked at boatyards, blacksmiths, and basket factories and people who operated freight boats that brought locally caught oysters into Fulton Market.[13] The firms that sold oysters to the wholesalers had to pay people to cut and haul stakes out to their leased beds; shovel oysters into and out of the water; transport seed oysters every year from as far away as the Chesapeake Bay; and shuck the harvested oysters, pack them, and process the shells into lime as an extra product from this industry.[14] Shucking an oyster means removing the meat of the creature from its shell. Even though the oysters often traveled only short distances, this early form of controlled aquaculture constituted a surprisingly extensive and complicated provisioning chain.

In eighteenth-century New York, oysters were both very common and very cheap—so much so that many poor families had little to eat but oysters and bread.[15] The booming early market for oysters caused a host of ecological problems long before Fulton Market even opened. These included overharvesting, the destruction of oyster beds (due to the use of shells for making lime), and the spread of predators.[16] In response, towns around New York City passed local restrictions on oystering because so many boats from other locations would come to the mouth of New York Harbor and just take whatever they could carry, since the beds were essentially public space.[17] During the 1820s, a group of Staten Island ship owners began buying immature seed oysters from New Jersey and the Chesapeake so that they could dump them on the depleted beds and restore the oyster supply. When the ships arrived from down south, the oysters were put onto dories, approximately fifty bushels in each one. Squares in the beds were staked out with poles, then five bushels at a time were distributed across each square.[18]

Even with a very limited capacity, oystermen in Staten Island still managed to grow an industry during the early nineteenth century. Many of the old families on Staten Island became wealthy by dominating this early trade. During the 1830s, a group of African American oystermen brought their families to the south side of Staten Island to start a settlement known as Sandy Ground, where they would cultivate oysters for decades.[19] The various underwater plots all around the island were separated by hemlock boundary poles. The business thrived most when demand for oysters was at its height, between 1860 and 1890.[20] However, because oysters grow very slowly, even seeded beds could only be exploited for a limited number of years. Firms therefore needed multiple oyster beds so that they could let the oysters at part of their holdings grow back. That usually took two or three years.[21] This requirement favored the consolidation of the industry over time because only large firms could rent or own enough beds to balance all the stages of growth among them.

The process of raising oysters on underwater beds resembled farming in the sense that oystermen planted young oysters on ground they cultivated and carefully tended their "crop." They left shells on the beds to give the young oysters a surface to which they could easily attach. They destroyed predators like starfish and a predatory sea snail called the oyster drill. They removed nuisance species like black mussels that competed with their oysters for food.[22] In the early years they did all this work even though the underwater poster beds were technically public lands. There was no system for leasing the bottom of the harbor anywhere around New York or New Jersey. Until that system came about, the ownership of beds was often challenged in court. Courts awarded bedders leases as a result of these disputes.[23] Eventually, both towns and the State of New York leased underwater tracts to oystermen to raise shellfish. "The land under water thus becomes practically the same as the land above water," explained a Fulton Market wholesaler in 1885, "a permanent property of the planter, and is worked just as upland is, to preserve it and yet get as much out of it as possible."[24] Starting in the 1880s, Fulton Market dealers began to enter the wholesale oyster business themselves, investing their earnings from the fishing business into an industry that was easier to horizontally integrate as long as the oysters came from nearby.

The first local oystermen used a rake and tongs to gather their oysters. The rake was used to scrape oysters off the bottom. The tongs was a means to bring them up.[25] The first oyster boats were long, deep canoes. Sometimes they had sails attached to them, but they always had to be big enough to hold the catch. By the time the nineteenth century began, rakes and tongs ranged from 7 to 24 feet in length—which tells you the depth of water in which they could be used.[26] Improvements in oyster boats—like sails—over the course of the nineteenth century greatly increased productivity. New York's oyster boat sloops debuted during the 1830s. They were designed to pull oyster dredges (essentially, an iron bar with chains attached to it that displaced oysters attached to the sea floor) and could hold between 100 and 600 bushels of oysters once those dredges crossed the oyster beds.[27] The advantage of using a dredge rather than a rake and tongs is that it could easily pull up an entire bed in just a few drags.[28] The problem with oyster dredges—especially tied to boats operating on steam power—was that they could rip out all the oysters from a bed and leave it essentially barren.[29] Local oyster boats transferred their cargo to larger sloops, which then sailed over to Manhattan to drop off their cargo in the locations where the wholesalers were located.[30]

Between harvesting and distribution, oysters from around New York were "floated" in waters close to the city. Floating involved soaking shucked oysters in fresh (or just less salty) water so they could absorb more water as it came in with the tide. This made them appear bigger to consumers than they otherwise would, which meant they fetched a higher price. Floating oysters dates back to the first days of the oyster industry around New York. However, the practice occurred in all the waters surrounding New York City and included oysters that started their lives in waters way outside of town. Floating became more common in 1865 after the invention of a wooden frame to hold the oysters.[31] Fulton Market dealers defended the practice on the grounds that it improved the taste of their product.[32] They also liked that the oysters absorbed more water and appeared fatter as a result. Floating also made it easier to transport the product, because they wouldn't dry out as quickly.[33] The practice continued until it became a health hazard after the turn of the twentieth century.

By the late 1880s, Fulton Market had become one of the two chief distribution points for oysters in New York City.[34] Not coincidentally, this is also when the oyster wholesalers began to consolidate. Essentially, they pooled their resources—boats, processing facilities, capital—to take advantage of the growing demand for oysters on an industrial scale. While many smaller loads still arrived by boat, oysters also began to arrive by train and by steamboat from New England.[35] The consolidation, industrialization, and vertical integration of the oyster industry were necessary to keep pace with demand, allowing the price of oysters to stay low so that most New Yorkers could afford them. Even in a bad year, the demand for oysters could be met by importing most of them from elsewhere.[36]

During good years, live oysters left the city in droves. Chesapeake oysters, brought to New York to grow on depleted local beds, were shipped out to places all over the country where people wanted fresh oysters. At first they departed in barrels, with layers of chipped ice separating them. By the 1870s, live oysters were sent out in tightly packed tin cans ranging from one pint to four quarts in size. The amount of oysters sold and the amount of tin needed to make cans made this an important auxiliary industry in its own right, although cans were most common in the New England retail trade.[37] The oyster export trade began in 1870, the first barrels of live oysters going to Great Britain. Because European oysters were smaller than American ones, the wholesalers were able to sell oysters that wouldn't have brought as much money had they been sold domestically.[38] Americans kept the biggest and juiciest oysters for domestic consumption, especially in New York.

Even after wholesale fish dealers entered the oyster business, Fulton Market was more famous as a place to consume bivalves than as a supply hub. Its oysters were a favorite of both tourists and locals. The market was also full of oyster stands that catered to people with less money and little time on their hands, including the people who worked in the market itself. For this reason, the oyster trade was vital to the evolution of the Fulton Fish Market even though it wasn't based there. It was part of its ambience; the culture of the place. The existence of the oyster trade helps explains why the fish market was so celebrated. Moreover, the change in the nature of the oyster business through Fulton Market mirrors the ups and downs experienced by the fish dealers whose trade was based there.

FIGURE 4.1 "Oyster stalls and lunch rooms at Fulton Market, Fulton Street, N.Y." Oyster stalls were the pre-eminent restaurants at Fulton Market during its retail heyday. While other markets in the city sold prepared food, the oyster stands and the ambience surrounding them attracted tourists from all over the world. From *Frank Leslie's Illustrated Weekly*.

Library of Congress, Washington, D.C.

Oysters were the main dish served at most of the restaurant spaces when Fulton Market was full of restaurants. "Oysters! oysters! oysters! oysters! everywhere in Fulton Market," exclaimed the *New York Herald* in 1869. "The rough exteriors of the sleeping bivalves are piled in outlines, both shapely and unshapely, at all quarters of this weather-beaten edifice. The first thing that lights upon the nostril as a person enters this place . . . is the strong smell of our hero."[39] By 1872, Fulton Market vendors sold more oysters than anywhere else in the city.[40] In 1877, *Frank Leslie's Illustrated Newspaper* reported that 50,000 oysters were consumed at Fulton Market every day: "Ladies and children turn down Fulton Street from a Broadway promenade; merchants and banker hasten there at noon; foreigners are escorted to the busy place almost before the dust of the city has settled upon them, and a myriad of people who daily cross the ferries between New York and Brooklyn, stop at their favorite dealer's, and either brace up with a raw or stew, or take a box of fries to the young and old folks at home."[41] Cooks had many possible ways to prepare oysters at home. If not served raw, they could be roasted, creamed, broiled, or fried.[42]

The small oyster stands inside Fulton Market were wood, and designed in an imitation Gothic style to attract attention. Little restaurants attached to these stalls sold a lot of oysters, mostly to the people who worked in and around the market.[43] A French visitor described one of these facilities:

> In front of the counters of these traders are large sheet iron furnaces, usually rectangular, about six feet long, six feet high and three feet wide. The upper part serving as a receptable [sic] for smoke is terminated by a pipe, which communicates with the outer air. The lower part, lined with bricks, holds a large quantity of coal, by means of which a hot fire is sustained. Upon the fire, and touching it, a gridiron is placed, and on this the mollusks are cooked, particularly the roasted oysters, for which Americans have a special predilection."[44]

Unfortunately, these stoves made the whole market stink of coal fumes. In 1873, Fulton Market's fruit dealers complained that these gases tainted all the meat and produce sold there after just a short time. By their account, during the busy season these grills burned thirty tons of coal or coke every week.[45]

At an oyster stand, a half dozen oysters meant seven. A dozen meant thirteen. While the product was invariably inexpensive, patrons inside Fulton Market had to pay more if they wanted to sit down while they ate.[46] Yet, as one guide later explained, "To insinuate oneself sideways into one of these little huts, and have set before you a bowl of stewed oysters, just off the stove, while the aproned man who served you stood with arms akimbo, and retailed the gossip of the moment with hearty good will and a genial admixture of slang, was a Bohemian experience which few old New Yorkers have not indulged."[47] A "Fulton Market Roast" was nine large oysters cooked in the shell, buttered and sprinkled with pepper. These were often served over a slice of toast.[48]

While raw oysters were often served in shells, shucking oysters was a necessity to prepare other oyster dishes. Shucking small bivalves might not sound like much, but a later government report describes why it could be very hard work:

> The process of shucking requires considerable skill and strength of hand and wrist. Various methods are employed and several different styles of opening knives. Some shuckers first break off the "bill" or tip of the shell with a small hammer, insert the knife into the opening thus made, and cut the large muscle holding the shell together. Others scorn such aid, since it takes longer, and by steady pressure force the knife between the shells at the tips of the side. A skilled shucker moves his hands so rapidly the eye can hardly follow the movements. A heavy mitten is worn on the left hand, which grasps the oyster, the shell being very sharp on the edges. A fair day's shucking is 10 to 12 gallons.[49]

Popular places in the market employed people to process their oysters right there in the market, in view of where people ate their dishes. Once shucked, oysters were sorted and graded for different purposes. Those that were bunched together were separated. The crooked or misshapen oysters went for eating raw in the less expensive local establishments. The cleanest-looking oysters generally went for shipment.[50]

On the high end of the trade, Dorlon's—a famous oyster house—was actually three different restaurants. This included two establishments in the market and one outside of it. George Dorlon ran a small oyster stand in the earliest days of Fulton Market. His two sons, Sydney and Alfred, opened up competing stands inside the market that sold both fish and oysters. One was called A. and P. Dorlon. The other was called Dorlon and Shaffer.[51] An 1866 guide to the city explained of this second restaurant, "Between the hours of six and twelve, P. M., no less than between three and four hundred ladies have been known to visit here, and partake of oysters, in one form or another. Of course, these were attended by cavaliers, swelling the number to no doubt, nearly a thousand."[52] Indeed, Dorlon and Shaffer was the more popular stand with women customers, which explains why it grew to dominate the trade inside the building, eventually employing thirty men to serve orders quickly, thereby increasing turnover. Both places were plainly furnished, and, taking advantage of their proximity to the wholesalers, served fish as well as oysters, although both were known for the latter.[53]

Another Dorlon's, unrelated to the best known enterprise in Fulton Market, opened on the south side of Madison Square in 1875.[54] It was started by a former executive of the Lord and Taylor department stores and a former cartman for Dorlon and Shaffer. "There is a place uptown . . . bearing the same name," explained that same *Frank Leslie's* reporter. "It is showy, gorgeous, highly-decorated, studded with large mirrors on the wall—in short, lacking in the very features that made the Fulton Market resort so cozy and comfortable."[55] But it did illustrate the extension of the supply line. Fish from Fulton Market originally had to be eaten at Fulton Market. By 1875, that was no longer the case. Alfred Dorlon, along with their brother Philetus (the P. in A. and P. Dorlon), went into the wholesale business, and they were among the first dealers to export oysters to Europe.[56]

Starting in the 1880s, thirty oyster barges, huge floating warehouses of oysters, sat on opposite sides of Manhattan, dispensing their product to a hungry public.[57] Every oyster dealer sold from barges, including some near Fulton Fish Market. Developed during the 1840s, the typical oyster barge was a two-story wooden vessel. These elaborately decorated boats eventually covered both the wholesale and retail trades. Oyster boats arrived with product at the back. The fronts of the barges looked like buildings or restaurants. Customers came up on foot to buy oysters for themselves or with their wagons to stock retail locations across the city.[58]

The barges near Fulton Market survived longer than the stands inside the market did. Perhaps that was because working off a barge avoided the cost of expensive Manhattan real estate. "Inside," explained the *New*

Yorker in 1928, these barges "are really small factories. Rows of men in rubber boots and rubber aprons stand at benches, armed each with a hammer, a tiny anvil and a hooked knife. With these they crack the shell, split it open, scoop out the oyster and drop it in a container, all with accomplished rapidity." Everything from offices for the business to ice machines to keep the oysters cold was located somewhere below deck.[59] These vessels were big enough that they could only be moved by towboats. For all intents and purposes, the barges found permanent homes all along the city's waterfront, staying in one place for years at a time.[60]

Besides the market and the barges, there were oyster stands throughout the city and at other markets, which allowed New Yorkers to just eat their oysters out on the street.[61] There were also, as of 1869, over 7,000 oyster saloons around the city that made considerable money selling bivalves and booze together.[62] At first these were mostly lower-class places that admitted both men and women and were way beyond the boundaries of respectability for most members of the middle class. Eventually, higher-end oyster saloons used fancy decoration and sex segregation to suggest an air of propriety.[63] Sometimes the clientele of an oyster saloon varied by time of day. Businessmen and tradesmen might pick up some oysters for breakfast. At lunch, the customers might be businessmen who needed a quick meal. By dinnertime, these all-day restaurants might take on the trappings of a traditional saloon. With more dishes on the menu, they attracted male and female customers alike.[64]

During the 1820s, an African American named Thomas Downing who had previously made a living harvesting oysters opened one of the nicest and best-known oyster saloons in the city. Despite his race, Downing served the city's elite and operated one of the few higher-class oyster establishments that were seen as acceptable for women to visit.[65] Charles Dickens, who dined at Downing's place, describes his opinion of New York's many oyster cellars in the 1842 notes of his visit to America. "At other downward flights of steps, are other lamps, marking the whereabouts of oyster cellars," he wrote, "pleasant retreats, say I: not only by reason of their wonderful cooking of oysters, pretty nigh as large as cheese-plates . . . but because of all kinds of eaters of fish, or flesh, or fowl, in these latitudes, the swallowers of oysters are not gregarious; but subduing themselves, as it were, to the nature of what they work in, and copying the coyness of the thing they eat, do sit apart in curtained boxes, and consort by twos, not by two hundreds."[66] Downing's success served as an important indicator that African Americans could achieve more by serving food rather than supplying it, since seafood wholesalers during this period were invariably white.

Oysters remained popular in the city for the rest of the century. The Scottish poet Charles Mackay described the range of oyster specialties he saw during a tour in the late 1850s this way: "Oysters pickled, stewed, baked, roasted, fried, and scolloped; oysters made into soups, patties, and puddings; oysters with condiments and without condiments; oysters for breakfast, dinner, and supper; oysters without stint or limit—fresh as the fresh air, and almost as abundant—are daily offered to the palates of the Manhattanese, and appreciated with all the gratitude which such a bounty of nature ought to inspire."[67] If oysters, as they say, are an acquired taste, it is easy to see how so many New Yorkers could have acquired it when they had so many options from which to choose when buying them. They also had options for how they wanted their oysters prepared. As Dickens noted, the signs that marked the city's oyster saloons tended to include the message, "OYSTERS IN EVERY STYLE," which rightfully suggests that there are many ways to eat an oyster besides popping it into your mouth raw.[68]

So what made the oysters in Fulton Market special? An appreciative correspondent wrote of Dorlon's oysters, "Simple as the cooking of an oyster may seem, there was a fascination in Dorlon's style which few could approach."[69] The businessmen who ran stands seem to have agreed with the general sentiment that simple was best. "Talking about cooking oysters," a dealer who worked with Dorlon and Shaffer explained in 1886, "I want to say right here that the plainer oysters are cooked the better. Oysters don't wear any frills." That dealer then went on to tell the reporter who was interviewing him every way he knew to cook an oyster in two long paragraphs of text. "Now you know almost as much about cooking oysters as I do myself," he concluded, "and I would like to have you let me know how you succeed as an oysterman." This unnamed oyster chef cited two African Americans as the inspiration for his basic approach to this particular food.[70] Overcooked oysters will shrink and become tough.[71] The oyster business may have been difficult, but the cooking remained simple.

"Dorlon's oyster saloon," wrote an angry correspondent in the *Omaha Bee* about both Dorlon's establishments, "are famous far and near and for no other reason than luck. They are small rooms right in the middle of the carcasses, vegetables and bad smells of the filthiest of markets. . . . The best oysters are well cooked in them, and that is all. Precisely the same material and cooking may be found in other oyster houses."[72] If the oysters in Fulton Market weren't really that different than in any other eatery in New York City, the only good explanation for their popularity would be the ambience in which they were served. After all, there couldn't have been a better place to eat a food that "tastes like the sea" than in a market surrounded by fish. Especially for tourists, eating oysters at Fulton Fish Market was a tasty way to see the underbelly of New York. It wouldn't be something to tell the folks back home about unless the oysters were somehow special too. "An oyster will taste like the taster expects," explained the legendary food writer M.F.K. Fisher in *Consider the Oyster*. Once Fulton Market's reputation was established, that was all it needed to stay famous.[73]

After the turn of the century, when oyster consumption dropped off and local oysters became more difficult to find, the Fulton Market dealers took over the entire oyster distribution system in the city. While fish shipments throughout the country started to contract, oysters from around New York remained a favorite across the country for decades. Fulton Market became more important for oysters at that point because its dealers had experience sourcing seafood from farther away, which was necessary to find oysters that weren't problematic because of the pollution of local beds. By this point in time, the market had grown to command the trade in fish for a large swath of the country. The dealers who dominated that market became forces in the industry, influencing consumers, the regulatory practices of the City of New York, and federal government policy.

5

THE OPERATION OF A WHOLESALE FISH MARKET

It can be very hard to follow or even differentiate between the many wholesale fish dealers that occupied the Fulton Fish Market. For example, the first occupant of Stands 8 and 9 in the new building occupied in 1847 was a firm called J. C. Comstock and Company. It became Comstock & Kingsland until 1868, when Comstock died. Then it became Kingsland & Comstock. In 1923, it became Bishop, Kronimus & Pannen when they bought the business. In 1924, Kronimus sold his interest to the other two partners, which made the name of the business Kronimus and Pannen.[1] Caleb Haley & Co., D. Haley & Co., and Eldred & Haley were different firms that operated at the same time. So were Amos G. Chesebro and Samuel Z. Chesebro.[2] Blackford & Company is easier to follow because it remained the biggest firm in Fulton Market for almost thirty years.

Born in 1839, Eugene G. Blackford began his career at age fourteen, working in an office and later under the department store magnate A. T. Stewart. In 1867, he was a bookkeeper at the firm of Middleton, Carman and Co. in Fulton Market when he decided to start a company of his own. He built up his business selling to stores, restaurants, and steamships. He built a freezing station in Canada for salmon. He was the exclusive agent in Fulton Market for Connecticut River shad. His company leased five miles of seashore near Montauk on Long Island to secure a regular supply of quality seafood for its stands.³ Blackford was also a pioneer among Fulton Market wholesalers in entering the oyster business. From 1879 to 1892, he served as one of four Commissioners of Fish and Fisheries for the State of New York. He was particularly well known for his interest in the science of fishes. One result of this interest was his work toward the establishment of New York State's Hatchery Station at Cold Spring Harbor on Long Island, where the state worked to breed fry for release into public waters.⁴ Red snapper, a Gulf of Mexico fish, debuted at Eugene Blackford's stands during the 1870s. Its scientific name, *Lutianis blackfordi*, stands as a monument both to its popularizer and to its connection to the Fulton Fish Market. When Blackford died in 1905 he merited an obituary in the journal *Science*.⁵

FIGURE 5.1 Eugene Blackford working his stand in Fulton Market, 1877.

The Miriam and Ira D. Wallach Division of Art, Prints and Photographs: Print Collection, The New York Public Library

More importantly for the significance of this institution, Blackford was also a consummate showman. He capitalized on everyone's basic fascination with the vast variety of fish in the sea. His offices at the market were filled with "fine oil paintings of our leading varieties of fish," reported the *Brooklyn Daily Eagle* in 1879. "The salmon, as large as life, lay in all its silvery brightness, tinted red about the gills and at the tail, upon a background of rich green; a Spanish mackerel was lying on the sand; a pair of brook trout, set across each other, were expiring in moss; pickerel and muscling stood together on the canvas, their accuracy verified by what was just outside the window."[6] Blackford's greatest contribution to the operation of Fulton Fish Market may have been his tendency to build out the spectacle of the place in order to attract visitors, something he probably learned while working at a department store. Down near his stands were Blackford's famous display tanks. "Large aquarium tanks are arranged tastefully on the stalls in which are exhibited specimens of live fish of various kinds," explained an 1892 guidebook for New York tourists. "Sometimes strange living fish may be seen here and have been brought from China and Japan. The King-i-o, the Gouramie, and the Paradise fish, and also the Axlotlee have all been shown at this place. Monstrous turtles, some of them weighing from four hundred to five hundred pounds each, may be seen here; in fact, everything that is caught in the ocean, river, lake or pond may be seen here at some time during the year."[7] The fame of Blackford's displays in his fish stalls helped make Blackford himself famous because he was not too proud to continue working at his stands even after his business grew large.[8]

Near his office was the assembly of fish-related items generally known as the Fulton Market Fish Museum, with an extensive library nearby. A correspondent for the *Providence* [Rhode Island] *Evening Telegraph* set the scene when visitors climbed the stairs to the second floor of the wholesale market:

> Opening the door the explorer finds himself in a large, handsome room, well-lighted and overcrowded with a multitude of objects. On the walls are great black wooden panels on which are mounted the oddities of the ocean world. On one is the parrot fish, a handsome but clumsy animal,

FIGURE 5.2 The Fulton Market Fish Museum. Every spring, Eugene Blackford ran a popular trout show in Fulton Market that included tanks of live fish, but the oddities depicted at the front of this engraving were available for viewing all year round. "The Trout Display at Fulton Market," by Daniel Beard. *Harper's Weekly*, April 15, 1882.

Author's collection

whose scales about the head, throat and breast are strongly outlined in scarlet, and the remainder of whose body is relieved in bright green.... On another panel is the head of a monstrous green turtle, at which first sight suggests the yawning jaws of a great dragon.... On the shelves and the tops of cases are sharks—the tigers of the sea.

Other sights in the room included an iguana, a "hell-bender," a "mud-puppy," and a lung-nosed gar, all stuffed for display.[9] Considering these, it comes as no surprise that Blackford played an important role in the founding of the New York Aquarium at Castle Clinton in Battery Park in 1896.[10] He was ahead of his time because he thought about fish in both scientific and environmental terms, and he tried to interest the people who visited the spectacles he created in the same things about fish that interested him. This reflected Blackford's pioneering idea of the wholesale fish business as an industry, rather than just a collection of individual firms.

In 1878, a fire severely damaged the original 1822 retail market building. Blackford led the effort to replace it. A report by the Board of Health after the fire offers an interesting picture of the state of the market at that time. It had "four drinking saloons, six restaurants, six coffee and cake shops, one book store, one boot and show store and a large number of variety stands," explains the *Times* summary of the report. "On the main floor much of the wood-work is saturated with animal matter, giving it an offensive odor, especially that portion of it devoted to the fish trade."[11] There was some debate over whether that building could be renovated. As the decision dragged out, recognizing that something new was coming there, the *Times* observed, "Everybody has called out a nuisance for more than a quarter of a century, and an unmitigated and unmitigatable nuisance it has been in many respects. Nevertheless, it has had its uses and advantages, its inconvenience, its rambling, awkward additions, its dinginess, its dreariness, its entire dissimilarity to anything that it ought to have been lent it a certain picturesqueness, such as squalor, combined with irregularity often insures."[12] It was finally demolished two years later.[13] All market activity took place in the 1869 building until the wholesalers and the city agreed on creating a replacement.

The new building that Blackford championed following that 1878 fire was both a wholesale fish building and a retail facility. This demonstrates how the retail business became less important to Fulton Market as time passed while the wholesale business became more so. A Victorian structure with dark red brick and terra cotta trim, it was larger than the previous building, having 218 stands instead of 163. There were fewer restaurants than before, in part because they all had to cook their food in a kitchen upstairs—open fires were banned from the first floor of the building in order to improve the air quality. A glass and steel roof over the interior court made the inside much brighter. It also had refrigerated storerooms and a telegraph office. This second new building in about fifteen years was a sign of the clout that wholesale fish dealers had developed at the market. It opened in April 1883, just six weeks before a new bridge over the East River changed Fulton Market and the neighborhood surrounding it forever.[14]

Despite new facilities every few decades and a provisioning chain that expanded and contracted at the same time, the Fulton Fish Market developed a reputation for never changing (even before Joseph Mitchell came along) that was never really justified. "Physical conditions have been changed somewhat by the advent of more modern market buildings," wrote the wholesale fish dealer George T. Moon in 1918, "but in the main the economical progress of the business in New York has not kept pace with the times nor with the advances made in business methods in other cities. The layout of our plant, the packing of our goods (we are still breaking ice by hand, etc.) are reminiscent of the past, and in this respect the injunction, 'Wake up Fulton Market!,' is apt at this time."[15] The market did "wake up," at least in the instance Moon cited. At least one ice-crushing machine appeared, presumably at some later date.[16] Even more important, the transition from retail food market to wholesale fish market was gradual but did more to change daily life at the Fulton Fish Market than any other change over the entire history of this facility. Fulton Market became a venue that the public primarily

FIGURE 5.2 *Fulton Market fish dealers—looking N. along South Str., New York City, U.S.A.* New York City New York, 1902.

[New York; London; Toronto-Canada; Ottawa-Kansas: Underwood & Underwood, Publishers] Photograph. https://www.loc.gov/item/2017659900/

identified with selling fish. Yet few observers recognized the significance of the difference between the two kinds of transactions.

Retail transactions were always very straightforward. For example, in 1860, *Frank Leslie's Illustrated Newspaper* wrote about Warren Leland, who procured all the food for the Metropolitan Hotel. He visited Fulton Market every morning, stopping at the butcher, the poultry dealer, and some vegetable stands before arriving at D. Manwaring and Co., "where the finny denizens of lake, brook and sea are displayed in all the varieties which the seasons afford." These included "The great sea bass from the eastern shores, the speckled trout from the streams of Vermont . . . the lake trout, the muskelonge, and the ravenous pike from the great lakes." On this day, Leland purchased 30 pounds of salt mackerel, 60 pounds of salt cod, 38 pounds of smoked salmon, 35 pounds of smelts, 34 pounds of striped bass, 36 pounds of trout, 200 crabs, 30 pounds of lobster, and a 213-pound green turtle.[17] He brought those back to the hotel and returned to the market to hit multiple stands whenever he needed new supplies. Retail operations had to be open during the daytime.

Most food sales occurred in the morning, and restaurants did most of their business during lunchtime or at night. The wholesale market, on the other hand, famously began in the middle of the night because the goal of the dealers was always to receive fish and sell it as soon as possible.

Along with its hours, the wholesale market required an entirely different way of doing business. It started with the way dealers procured the fish that they sold. Until 1936, wholesale fish dealers sold all of their fish on commission. Unless they were themselves involved in procuring fish, they literally risked nothing when doing business. In 1936, under pressure from the U.S. Attorney's Office in New York for possibly violating racketeering laws, every dealer in the Fulton Fish Market stopped selling goods on commission and instead purchased their product from fishermen outright.[18] This forced the wholesalers to assume at least some risk, which probably made them at least a bit more efficient. However, the fishermen still had no assurances that they would receive a fair price from the notoriously collusive wholesale market dealers there. "It doesn't make any difference which [firm] it is; they all cheat us," complained a fisherman in the *Daily Worker*:

> They tell us that all they will be able to get for the fish is 6 cents a pound or 7 cents. They give us what they want to. Then, after they give us this small amount, they go out and get 14 cents a pound for it. When the fishermen come in to the market, Haley or Hansen or the other companies tell us they are going to pay us 6 to 9 cents a pound. They take the fish and, when we return for our money, the company tells us: "All you'll get is 3 cents a pound. If you do not like it then go to hell."[19]

This kind of exploitation, along with the costs imposed by organized crime, improvements in refrigeration, and the increased competition from wholesalers in other cities, explain why the fishing industry moved steadily out of New York over the course of the twentieth century.

The distrust between wholesalers and fishermen stood in stark contrast to the warm relationship between wholesalers and their regular clients. Buyers and sellers knew each other well. Sometimes the deals were worked out even before the buyers arrived. Wholesale buyers came to

Fulton Market just to purchase fish, and nothing else. If they didn't already have a contract with a wholesaler, the buyer could see what every dealer had and, if they wanted some of any particular product, try to come to an agreement in order to buy it. Although the amount of seafood that Warren Leland brought back to his hotel may seem like a lot, wholesale dealers sold fish in larger amounts because the deals they arranged often lasted more than one day. A Bureau of Fisheries report from 1926 described a typical wholesale deal:

> Each container is labeled with a mark or number before being placed in the section of the stand or department handling that particular variety of goods. The retailer or jobber visits the stalls of the various firms and selects the products he wishes to purchase. When a sale is made the salesman calls to the clerk stationed in the sales office the mark or number of the consignment, the weight, price and name of the purchaser. If the goods are sold to a dealer doing business in the metropolitan area delivery is made by the wholesaler to the customer's truck, or if sold to an out-of-town customer the package is prepared for shipment by express or freight, as the case may be.[20]

If the dealer was big enough to own or invest in their own fishing boats, they could sell the same fish at even lower rates because there were fewer steps in the transaction. This cut down waste by making sure that more fish was spoken for at the end of every day.

The gradual growth in the wholesale seafood trade was an extraordinary change in the way that the Fulton Fish Market did business. Other, smaller changes also happened during the late nineteenth and early twentieth century. For example, 1921 saw the formation of the first fish forwarding company, which bought up the catch of various fishermen and sent it all to New York City via train from places like Boston or Baltimore together. This saved on transportation costs and allowed dealers to get paid faster since it meant that fish arrived at the market faster.[21] That fish stayed at the market less time since it was often spoken for even before it arrived, thereby bypassing the wholesale marketplace but not the market. This improvement was made possible by the growth of telephone service. "Outward movement of fish from the market to the retailer and consumer is governed largely by telephoned orders from buyers in every section of the country east of Chicago and St. Louis," reported the *East Hampton Star* in 1949. "Also, to keep the vital food supply moving, large forces of salesmen representing the Fulton Market wholesalers use the long distance telephone constantly in contacts with out-of-town buyers."[22] The telephone marked the beginning of modern logistics, allowing knowledge to spread from one link to another within the provisioning chain.

While there were traditions that created the appearance of a place that never changed, in reality the wholesale fish dealers made many small adjustments in order to keep business at the market going. Even though the volume of fish going through the market dropped over the course of the twentieth century, they did well enough to ward off the move to the Bronx until the beginning of the twenty-first century. Because reporters and other outside observers had little interest in or understanding of how a wholesale market operated, they continually repeated a story about an institution that was essentially frozen in place. There was a rhythm to how the market operated across the decades, but to suggest that it couldn't cope with change or never changed itself is simply incorrect.

The Fulton Fish Market had to receive fish in order to sell them. Around the turn of the twentieth century, most of the supply of fish arrived between April and September.[23] With little cold storage available, the wholesalers had to sell everything very quickly. "The shrewd commission merchant was the man who had a knowledge in advance of either glut or scarcity," remembered Al Smith in his autobiography.[24] There were busy days, and days when little or no fish landed so very little could be sold. Thursdays were always busy because of the Catholic practice of eating fish on Fridays. One of the great benefits for the wholesalers of working on consignment was that if they couldn't sell their fish, they were out nothing. Only the fishermen or the other dealer who placed that fish with them suffered the economic damage.

FIGURE 5.4 "Interior of the Fish Market," *Leslie's Monthly Magazine*, August 11, 1877.

The Miriam and Ira D. Wallach Division of Art, Prints and Photographs: Print Collection, The New York Public Library.

Only in the market's first days did fishermen sell their catch right off the boat. Boats gradually gave way to trains and trucks when those became the usual method of delivering fish during the 1920s. Since there was no rail link at the Fulton Fish Market, even the fish that arrived in New York by train had to travel to South Street by truck or by barge. Whether from boats or trucks, unloading began around midnight each evening. Unloaders unpacked tons of seafood from crates, cartons, or barrels and distributed it to stands around the market. Once it arrived, market workers had to prepare the catch for sale. During most of the market's long history, wholesale transactions began at 3 a.m. on Mondays and 4 a.m. Tuesdays through Fridays. Journeymen brought fish that had been sold to the correct customer's van. Loaders directed traffic around the market. Sales ended around 9 a.m.[25]

Although sales were supposed to be restricted to stalls, there was activity in front of buildings, in open spaces, and all around the neighborhood too. Salesmen would stand around surrounded by stacks of crates (later boxes) of dead fish, some opened so that buyers could see and sample the product. Workmen at the market used bailing hooks to move wooden fish boxes and sometimes to spear either live or dead fish. The salesmen wore white dusters, heavy rubber boots, and straw hats. Winters were particularly bad at Fulton Market because so much activity took place outside. Even the work areas of most buildings had lots of openings. "You face a vista of men who at this season look more like Esquimaux [eskimos] than like 'New Yorkers,'" explained a magazine writer who visited in the fall of 1895. "They are done up in rubber boots because the floor is a sea of mud and water, and in sweaters and overcoats and caps because it is cold so near the water, and they are handling iced goods for hours at a stretch."[26] The men worked almost entirely at night, after all.

No auctions occurred at the Fulton Fish Market. Buyers would bargain individually with sellers. The wholesalers would have a price in mind at the beginning of the day for each type of fish and would try to get as close to that as they could.[27] Prices could fluctuate wildly over the course of the week as some fish spoiled or new supplies arrived off a boat or via train. The fact that many buyers purchased their fish on regular schedules, which the salesmen knew, helped smooth out the ups and downs for individual species. There may have been competition for particular lots, but never an auction in the traditional sense of that word.[28] Because plenty of fish bypassed this bargaining process entirely, the market for some species wasn't even particularly competitive.

This explains how the salesmen usually got the best of the people who bought their product. They knew far more about supply and demand than their buyers did. The *New York Evening World* described the advantages of wholesale dealers bluntly in 1919: "Each [wholesale] dealer has his separate stall, his clientele of retail dealers and his list of fishermen for whom he acts as middleman, receiving 12 1/2 per cent. commission on the sales. There is intense competition, and although this is a fine thing for the fisherman who gets his 87 1/2 per cent., it is a poor thing for the retailer who finds prices varying widely from minute to minute."²⁹ Similarly, fishermen got nothing for the fish that didn't sell at all (at least until the wholesalers started buying their catch outright in 1936). While there were risks involved for both sides of the initial transaction between fishermen and middleman, the wholesalers usually took home most of the profits.

Despite this situation, the wholesalers and their regular customers built bonds of trust based upon their familiarity with one another. Consider the very real possibility of fish fraud. In 1938, the *New Yorker* writer Russell Maloney visited the Fulton Fish Market. A buyer he called "Nick" told him that pollack often became bluefish and that fish he called "permit" often became pompano. "They were, he was careful to explain, impostors only after they left Fulton Market," wrote Maloney. "There they are properly labeled and fairly priced; it is only the undiscriminating restaurant and retail-market trade that can be taken in by such frauds."³⁰ Fish wholesalers didn't rip off their regular buyers because they wanted their business again. Whether retailers deceived indiscriminate customers was another matter entirely. Relationships based upon trust proved particularly important for selling a product like this one, for which the quality is difficult to discern with the naked eye and can go bad so easily. Who you knew mattered. What you knew about fish mattered. When you bought your fish mattered.

After a public sale, the salesmen called out a unique code to their bookkeepers in the rooms adjoining the marketplace to record the results. Here is the transcription of those sounds that one reporter wrote up after a visit to the market in 1874: "'One hundred fifty live cod, at four, Smeely; ninety-six haddock, Podgers; ninety-eight smelt, Quiggles; eighty-seven cod, Slurker; sixty-six white fish, Pennyroyal; eighteen mackerel, Moses; hundred and thirty-nine Jennie-cod, Fubbles; two hundred and eleven halibut, Blackford'—all these calls sounding from the most lusty of voices of powerful young men, rang and echoed through the market."³¹ It would not have sounded much different a hundred years later. Al Smith later explained the system to a reporter from the Associated Press: "We had a code for our prices, based on words up to ten letters. We would whisper the price to the customer then call out the price in code for the clerk, so the other commission men wouldn't know what price we were selling at."³² Nonetheless, it wasn't hard to figure out any firm's codes when they were heard every morning.³³ The real purpose was to prevent other customers from knowing the price at which fish was sold, since price discrimination was a common practice at the market.³⁴ This is why the dealers called out prices in shillings, a quirk that many visitors to the market found completely unexplainable.

Fish changed hands all through the provisioning chain, but at some point it had to be processed. Sometimes that simply meant killing the fish. Sometimes processing took much more work. Much of the fish sold at the Fulton Fish Market was processed by the companies that sold it. "After the boats unload," explained a writer from the Works Progress Administration in 1940, "an army begins wielding knives, cleavers, clippers, and scalers, preparing tons of fish, arranging them on beds of cracked ice for rush delivery to jobbers and retailers. The action along the waterfront is fast and furious since fish is the most perishable of commodities."³⁵ That's why wholesalers struck bargains with their most loyal customers before the fish even arrived, guaranteeing certain amounts of the most popular and common on a regular basis. It was the special catch—the species that appeared irregularly in the market or fetched a particularly high price—that got placed on display for buyers to compete over.

During the late-nineteenth and early twentieth centuries, a lot of the fish that went through Fulton Fish Market went to retail fish dealers who operated stores selling exclusively fish in neighborhoods throughout the city. As late as 1939, retail fish store owners did 75 percent of their business on two days a week, Thursday and Friday, to meet their customers' demand for fish on Friday.³⁶ They brought wagons to Lower Manhattan early in the morning to transport the fish they bought to their stores. A fish retailer likely took his wagon down to Fulton Market just once a week to supply the greatest number of customers with the freshest fish possible.³⁷ Together, these were the city's local fish distribution system for the Fulton Fish Market in those days. Over time, trucks gradually replaced the wagons. If a buyer was from out of town, the dealer would ship their fish directly to them, usually by rail.[38] "Almost all of the fish sold in the city comes through the Fulton Fish Market," wrote the *New York Evening World* in 1919. "The market is a great distributing center to the thousands of small dealers."

Gradually, fish stores faced competition from butcher shops, and later supermarkets. Neighborhoods that didn't have fish stores depended upon other stores—often butcher shops—that also carried fish. Butchers only had fish one day a week as an accommodation for customers who couldn't eat meat on Fridays, but as this was the most important day of the week for fish retailers, that competition hurt. Since butchers knew little about handling or storing fish, it tasted subpar and hurt the reputation of fish with consumers. Supermarkets, which became popular during the 1920s and 1930s, often used fish as a loss leader. They bought unpopular fish in huge quantities and undersold the expensive popular species that retailers tended to feature.[39]

Fish retailers also had to fight off competition from peddlers. Toward the end of the morning, peddlers would circle the fish dealers looking for bargains (at least until Mayor Fiorello La Guardia banned pushcart vendors in 1938).[40] When dealers lacked wholesale customers to buy their fish or storage facilities to keep them frozen, they sold what remained to the peddlers. "We would be ruined without those fellows," one dealer told a visitor in 1871. "The pedlars [sic] are our safety valve. They are always running around to see who is worse 'stuck.'"[41] After retail dealers arrived in large numbers later in the century, they formed an effective backup system. "No matter how large the supply happens to be, there is never a glut in the market," wrote the *New York Evening World* in 1888. "What cannot be sold in the regular way can be disposed of to peddlers. If prices are low they will buy any quantity, and if high the retail dealers will buy."[42] Peddlers were the last steps in the distribution system for food and other products to poor New Yorkers around the turn of the twentieth century, but even this did not prevent substantial waste at the market as the supply of fish increased over time.

The peddlers' goal was to pay as little as possible for the fish that more established businesses didn't want, so that they could sell it to less-discriminatory household customers before it spoiled completely. Without the need to pay rent on storefronts, they could set their prices a lot lower than those at retail stores. Despite the presence of peddlers, for much of the twentieth century, at the end of each day the market was full of wasted fish that had to be cleaned up one way or another, either by disposal or perhaps by the stray cats that famously frequented the market looking for unwanted parts of fish.[43] There were two kinds of waste. First, whole fish went bad because they couldn't be sold at all. Second, until canned cat food manufacturers became important customers in the mid-twentieth century, the unwanted parts of fish processed at the market were also wasted. Waste in any food chain is a sign of inefficiency. That indicated the possibility that markets elsewhere could displace the Fulton Market's role by managing it better.

The city stationed two sanitary inspectors at the market to make sure that the peddlers' bargains were still safe to eat.[44] "The dealers within this market have to be watched very carefully because some would not hesitate in selling spoiled fish should the opportunity present itself," explained the City Health Department in 1920.[45] The implication was that wholesale dealers were willing to cheat peddlers and that customers who got bad fish would have blamed their retailer rather than the wholesaler, whom they didn't know. However, since fish peddlers had to have at least some of the same knowledge about their product that the wholesalers did, it seems likely that there were simply different definitions of what the minimum acceptable level of freshness for fish was. Wholesalers were separated from consumers on the provisioning chain, but the bonds of trust between wholesalers and their most important customers were essential to keeping the entire market going.

Creating bonds of trust at the end of the provisioning chain was easier when the retailer and the customer shared the same ethnicity. In 1903, the *Tribune* described an unnamed German butcher who picked up fish specifically for his customers on Fridays. They also told the story of "Carp Carrie," a mysterious buyer who apparently ran a fish and meat shop on

Essex Street in the Lower East Side.[46] The *New York Evening Post* broke down the ethnic preferences for particular fish in 1902: "The Irish like cod, the English flounder, Americans bluefish, the Germans smelts, while the Russian Jews are extremely fond of carp, especially the variety which has the brilliantly colored scale plates."[47] Jewish fish stores, an important category of the trade during the interwar years, tended to feature freshwater fish.[48] When the first shrimp arrived at Fulton Market around 1918, only the Chinese would buy it.[49] Each ethnicity had their own favorite fish, and Fulton Market had wholesalers who carried it for them. When the tastes overlapped, the wholesalers could sell that fish to every group that wanted it. If the wholesalers got a lot of a particular fish by serving a group, they could try to market it to native-born Americans too.

When the Fulton Fish Market moved from retail and wholesale to an exclusively wholesale market, its role in the fish provisioning chain changed too. It moved from the center of every chain in the whole country into a dominant position simply for greater New York.[50] New York still received fish from increasingly distant waters, but it sent far less fish out than it had in the nineteenth century because the provisioning system gradually decentralized. New York City still had the power to determine trends because of its importance as a city, but the wholesalers who survived did so in part by expanding their business, creating distribution hubs that didn't include the Fulton Fish Market. This practice put increased pressure on fish stocks all over the world.

6

FISHERIES AND THE FISH MARKET

As the volume of seafood going through Fulton Market increased, the effects of overfishing and other forms of environmental destruction gradually became obvious. This was particularly true for common food fish found close to the city. Overfishing, a problem that began in the waters off New England even before the market opened, was not caused by the existence of Fulton Market, but having a place where someone could buy just about any sea creature only exacerbated the harm. The various forms of industrial fishing that came to be known as trawling never would have developed unless places like Fulton Market offered them an outlet to sell their ever-increasing catches. By the late twentieth century, the decreased supply of fish in the market reflected the depletion of waters across the world even as the variety of fish available actually grew. Conservation laws passed in response to these obvious shortages affected life in the market. So did changes in price, which were a reflection of the available supply of any kind of seafood. Therefore, understanding the systems that ran through Fulton Fish Market can help us understand the interaction between humans and their environment, especially by examining the market for a few products sold in high volumes over many decades, like menhaden.

The menhaden is a small, oily, bony fish that few people ate during the late nineteenth century and almost nobody eats now. Sometimes called the mossbunker, bunker, pogy (as opposed to porgy, which is a different fish), or alewife, the menhaden is the fish that served as fertilizer for both Native Americans and the earliest American settlers, who needed it for their crops to survive. They are also important to the ecological health of the ocean because they are filter fishes that consume algae. Because they travel in large schools, they became a prime target for early industrialized fishermen who could get lots of fish with little effort when they used particularly efficient new technologies.[1]

Although almost nobody actually ate these fish, the menhaden had a significant effect upon the history of Fulton Market at first because of its popularity as bait. "Twenty five years ago and before it was thought to be of very small value," wrote George Brown Goode in his epic, *A History of the Menhaden*, published in 1880. However, "As a bait-fish this is found to excel all others. For many years, much the greater share of all our mackerel has been caught by its aid, while our cod and halibut fleet use it, rather than any other fish when it can be procured."[2] When fishing boats could get enough menhaden, the results could be very lucrative. In 1878, the *New York Times* reported on a smack fishing for cod off Nantucket that returned from its 11-day trip with 1,450 cod, which altogether weighed about 20,000 pounds.[3] At first, menhaden was used as bait on trawling lines—applied to hooks on an individual basis through the hand labor of fishermen. Later, dories carried menhaden that had been pulverized by machines on shore. This was cast directly into the water to attract schools of bigger fish into large nets.[4] In fact, menhaden was the first fish that were themselves caught in large nets.

Once menhaden became popular as bait, people began to wonder how long the supply would survive. Those fears only grew when the menhaden fishery moved from being primarily used for bait to primarily being used for menhaden oil, which became a popular source ingredient of a variety of industrial products, from "extract of beef" to feed for domestic animals.[5] Echoing the beef processing arrangements then occurring in Chicago, the menhaden factories dried the fish carcasses left over from the oil extraction process and sold what remained to farmers as fertilizer.[6] In 1878, there were seventy-eight menhaden oil factories in the United States.[7] Many were on Long Island. Menhaden steamers would take the fish in large numbers and deposit them directly on piers built near those factories, bypassing Fulton Market completely. The state of the menhaden didn't matter if you were turning them into oil, so there was no ice on the boats and no refrigeration on shore. As a result, the smell of these operations was absolutely awful.[8]

The best years for the menhaden industry were between 1880 and 1890, when the fish first became an important source of oil and before other oils began to compete with it. Despite widespread fears of overfishing, the menhaden remains an important ingredient in industrial products to this day. The menhaden schools did become severely depleted in northern waters. In 1879, they disappeared from the coast of Maine entirely. The depletion of the menhaden fishery from the waters off Long Island north through the rest of New England meant that the many factories that had opened to process the fish had to shut down. By the 1890s, the center of the menhaden industry had moved from the waters off Long Island to those off Virginia as the fish became scarce farther north.[9] That explains why menhaden fishermen moved into places like the Gulf of Mexico in order to supply both fishermen and the menhaden factories throughout the country.

Observers were also worried about the effect the disappearance of the menhaden would have on other fish since the steadily increasing scale of the hunt for menhaden meant that the vessels looking for them caught food fish as bycatch. As was the case with young shad, food fish ate menhaden, so its loss would have hastened their demise too. Fulton Market dealers wanted to sell whatever food fish they could get, and often did when the menhaden vessels landed in Fulton Market to unload the fish that weren't wanted at the oil factories. However, when the United States Fish Commission made an extensive study of this issue in 1894, it found that only .33 percent of the menhaden trawls were bycatch. This helped quiet efforts to shut the trawling down.[10]

The longest-lasting impact of menhaden fishing on Fulton Market came when the owners of vessels that caught them began to adapt those ships to catch schools of food fish using similar methods. In July 1882, the *New York Times* reported that the first menhaden steamer was adapted so that it could target schools of mackerel instead. The next month, the *Times* discovered that there were actually somewhere between three and five such vessels pursuing mackerel full time. In 1885, a Jersey shore fisherman who casted from the beach complained, "I have seen them run around after a school of moss bunkers . . . and I have counted upward of fifty of the pirates in one day. They are ruining the fish along the shore and ruining us." These vessels came to be known as trawlers. Like menhaden, the mackerel caught en masse by these ships were used for oil rather than for eating. The supply of mackerel off of New England crashed.[11] So did the supply of lobster and halibut. In response, dealers began importing fish caught by Europeans for the first time.[12]

While the use of menhaden as bait threatened food fishes of all kinds, the development of trawling affected all fish because long lines and especially large nets scraped along the ocean floor picked up species indiscriminately. The concerns about the effects of trawling on fish stocks came as soon as long trawl lines first appeared on American boats in the North Atlantic during the mid-1850s.[13] As long line fishing took off, it tended to be practiced by two men traveling in a dory launched from a larger fishing boat. The more dories on a fishing boat, the larger the area it could cover at the same time. This was generally called tub trawling because the long baited lines were kept in a tub until they were dropped into the sea to catch fish.[14] By the mid-1870s, a single vessel could work twenty-five trawls at once with just four to seven men.[15]

Trawling, which had begun during the 1850s, had expanded both in amount and in impact per fishing boat by the 1880s. A cod trawler at that time might have 25,000 to 35,000 hooks on a single line, curled up in a tub and ready to deploy when a smack reached the fishing grounds.[16] The result was predictable, not just for menhaden but for nearly all the fish sold at Fulton Market. "The idea that some people have that fish are inexhaustible is wrong," complained a former New York fisherman in 1907. "My experience proves that shad and weakfish and bluefish, moss bunkers and Spanish mackerel and a host of other fish that were plentiful half a century ago in the water around New York have just been fished out."[17] By 1927, every boat still landing at Fulton Market had been mechanized, which only made them more efficient.[18] Since the wholesale dealers there depended on selling fish for their livelihoods, they adjusted easily to selling more of it.

Trawling with long lines gradually gave way to trawling with large nets. Some were set by small boats near shores and left in place. Others were carried by larger boats farther out to sea. The first efforts to catch entire schools of fish this way began as far back as 1853. Although this kind of trawling could only be used on fish that gathered in schools near the surface, it was an important first step toward what can clearly be seen in retrospect as industrial fishing.[19] Before long, nets were literally dragged along the sea bottom to sweep up whatever fish were there, even if the crew only sought one in particular. By 1883, one such net used off the coast of North Carolina could sweep "1,000 acres of the bottom. Net and rope are four miles long; the net being 2,480 long by 9 yards deep."[20] Taking thousands upon thousands of fish out of the water at one time with no regard for whether the stocks would ever be able to replenish themselves had an extraordinary impact on every part of the Atlantic Ocean. However, it seldom affected the supply of fish available in the Fulton Fish Market because fishermen could still easily move to other locations where particular fish remained, assuming there were large supplies available in the first place.

What made this change so important to fishing in general was that instead of fish coming to the fishermen, the fishermen scooped up all the fish in an area at once.[21] This description from the *New York Sun* of catching sea bass off North Carolina illustrates the change well. "The sea bass grounds are first strewn with stosh [ground bait, usually menhaden]," the paper explained,

> which slowly drops down through the water, attracting the fish. Then additional quantities of stosh are thrown on the water, and while the fish are near the surface, devouring the food with avidity, a purse seine is lowered overboard around them, from two boats rowed in opposite directions as rapidly as possible. When the fish are surrounded and the boats are brought together, a 500-pound weight is hitched onto the

purse line ... This weight is dropped overboard and purses the seine together, thus corralling the fish before they have time to dive down into the haunts from which they have been lured.

A purse seine is a very large net that can be snapped closed instantly at the top.[22] Eventually, steam-powered hoists would be able to bring the nets back on ship, making the process even more efficient.[23] Ships that fished this way came to be known as trawlers. The name arose because of the practice of trawling the bottom of the ocean with large cone-shaped nets that take in all the fish in their wake.[24] The word morphed again later as both technology and practices evolved.

Improvements in trawling instituted between the 1880s and 1920s made this kind of fishing much more efficient. Nets became easier to open and close and to run across the bottom of the ocean so as to pick up more fish. The size of the nets grew, and steam power allowed for trawlers to go farther out to sea and deplete areas that had not been fished by smaller boats.[25] Steam power also helped fishermen "run down" a school of whatever fish they sought after it had been sighted.[26] The technical changes in the nets and their operation allowed boats to employ much bigger nets without snagging and therefore catch even more fish at one time.[27] American scientists condemned these improved trawling methods for depleting fisheries, but there was not enough opposition to get the federal government to ban these practices.[28]

As long as Fulton Market wholesalers were selling all of their fish on consignment, they took on very little risk. They made money when large catches were sold by selling more than before, even at a comparatively lower price than before. "Flatfish and other varieties have been turned into Fulton market so fast during the last few years that the bottom has been knocked right out of the price," complained a Connecticut paper that sympathized with the fishermen there, the same year that federal legislation was being considered. "The fish broker at New York or at Boston is the only person who makes on the present arrangement. By icing the fish as fast as they are received he can control the price to the retailer to a considerable extent and assure himself a profit in many cases of several hundred percent."[29] These practices set the stage for the development of competition as soon as selling fish on a wholesale basis became possible elsewhere. Perishable fish had to be sold quickly, but once refrigeration, and later freezing, made them more like other commodities, the New York dealers lost what had once been their near-monopoly control over the product.

Even absent competition, the technologies and practices at the Fulton Fish Market had to change in order to absorb all of this fish. Rudimentary cold storage facilities began to appear around this time. However, the problem that all fish dealers faced was that large catches arrived at the market at irregular intervals. Most of the supply of fish arrived between April and September.[30] When any particular vessel found a large school of any fish, that alone could glut the market. If a number of different vessels found the same huge school of a popular fish, then trouble selling all that fish was virtually assured. With little cold storage around Fulton Market (or prices so low that keeping some types of fish in cold storage was uneconomical), a lot of fish ended up wasted, or maybe going for fertilizer. This had ramifications across the marketplace for fresh fish of all kinds.

In 1881, for example, Fulton Market had to deal with 200,000 pounds of weakfish arriving at the same time because three different menhaden boats mistook them for the fish they wanted. The Fulton Market dealers took all they could handle, and the price dropped accordingly. From two dollars, the price dropped quickly to twenty-five cents per fish. Then the smacks that brought the fish in from those menhaden dealers started giving it away. After that, the extra fish were dropped off at the same city island where dead horse carcasses went so that they could be turned into fertilizer.[31] In a different instance, the *New York Sun* reported that all the cheap mackerel decreased demand for every other fish available that day. The paper noted, "There was an air of gloom among all the fish traders in the market, and the fishermen themselves, instead of being elated over their phenomenal catch, were obviously depressed and unhappy. The possibility of getting too much of a good thing was being brought painfully home to them."[32] Al Smith remembered that when he worked at the market during the 1890s he could take home all the fish he wanted.[33] This suggests that surplus was a persistent problem during this era.

Occasionally these situations worked out more favorably. In November 1897, a single trawler temporarily glutted the market for cod. That steamer happened to be named *The Fulton Market*, and it arrived there with 50,000 pounds of fish aboard. In this case, the market had been warned in advance that the huge load was coming. Facing a glut of cod already, the wholesalers turned to Boston to buy some of the oversupply. The price dropped from 90 cents per hundred weight to 40 cents per hundred weight as they called around and took orders from Chicago, St. Louis, Cincinnati, Pittsburgh, Cleveland, Buffalo, Rochester, Albany, Newark, and Trenton. The *New York Sun* concluded, "The quickness with which this load that the Fulton Market brought was disposed of is a sample of the way that an over supply of any fish is handled in this market. Notwithstanding the tremendous run of cod for two weeks, there has been no glut, there have been no fish wasted."[34] This was the purpose of not just the Fulton Fish Market but all fish markets. They existed to even out a naturally uneven supply, by selling all the fish they couldn't store by whatever means necessary.

There were periodic efforts on both the state and federal level to limit trawling to control its devastating effects of depleted fish stocks.[35] The wholesale dealers seldom got behind them.[36] Since most of the fish at the Fulton Fish Market were sold on a commission basis until 1936, dealers made money on each sale, even when prices were low. They also made money on each sale when prices were high. They lost nothing if the fish they displayed on behalf of fishermen didn't sell at all. As long as there was any amount of a particular fish available, they made money. "While some individual fisheries, such as the mackerel industry, faced catastrophic losses," writes the historian W. Jeffrey Bolster in his book *The Mortal Sea*, "most fisheries, such as those for lobster and menhaden, found compensation for ecological depletion through rising prices. The market masked the mess."[37] Whether or not wholesalers understood what they were doing, the length and complexity of the provisioning chains they brought into existence obscured the growing extent of overfishing to both consumers and potential regulators. That's why industry leaders had to seek new supply sources for popular fish.

Over the course of the late nineteenth century, an increasing amount of that fish was bred in hatcheries. This was particularly true for trout. There was a seemingly infinite variety in lakes and streams throughout the United States. Referring to the kinds of trout close to the city, George Brown Goode wrote in 1887, "Every lake of Northern New York and New England has its own variety, which the local angler stoutly maintains to be a different species from that found in the next township."[38] All those trout were not enough for the nation's anglers. Fish culture involves the artificial fertilization of fish eggs in hatcheries, and the resulting fry being released in the wild. The first attempts at artificial fish breeding in the United States were focused on trout when a Cleveland area doctor bred eastern brook trout beginning in 1853. After the Civil War, the U.S. government began to fund the first efforts at large-scale fish culture, which focused on shad and salmon. However, the first really successful effort on the East Coast involved the raising of rainbow trout imported from California during the late 1870s.[39] Trout fishing appealed to sportsmen, not fish dealers, but this eventually led to more successes at breeding more commercial fish.

Eugene Blackford ran the fish exhibit at the 1876 Centennial Exhibition in Philadelphia.[40] Starting in the late 1870s, he organized a yearly exhibition at Fulton Market devoted to trout. It opened on April 1, the first day of trout season, and attracted thousands of visitors. The *New York Sun*'s report on the 1885 show gives some idea why so many sport fishermen came. "Mr. Blackford's display was of remarkable interest," its reporter wrote, "considering that winter had maintained its icy grip on stream and pond almost up to the opening date. Many live specimens flashed their bright colors inside large glass tanks. Heaps of brook trout were piled on marble slabs frozen and rigid, as if they had been scooped in that condition from mountain streams."[41] Trout did show up in Fulton Fish Market, but they were never an important commercial catch.[42] Introducing more trout to inland streams proved a good way to distract sport fishermen from the depletion of other, more commercial species.

FIGURE 6.1 Eugene G. Blackford business card, printed by Stahl & Clauss, 1881. Blackford advertised his annual trout show widely to draw ordinary New Yorkers to what had already become primarily a wholesale fish market.

South Street Seaport Museum, New York, NY, gift of Peter Neill

Both sportsmen and wholesalers counted on fish culture as a way to restore fisheries that were depleted by dams, pollution, or overfishing. There would be much greater tension between sportsmen and commercial fishermen over the course of the twentieth century, but the two groups were allies as long as hatcheries were focused on restoration.

Spencer Baird began working for the Smithsonian in 1850, after a brief career as a professor of natural history at Dickinson College. For twenty years, he collected fish from the waters off New England and heard many complaints directly from fishermen about depleted fisheries. In 1870, with $100 from the Smithsonian and a vessel owned by the Treasury Department, he began investigating the problem himself. In 1871, Congress created the United States Commission of Fish and Fisheries in order to investigate depleted fisheries throughout the country and put Baird in charge. Because of lack of funding, Baird operated the commission out of his home.[43] Since conservation wasn't a particularly popular solution with either fishermen or consumers, Baird made the creation of hatcheries to breed fish and restock them in their wild habitats a popular and successful way to combat the problem.[44] Much of that work focused on trout rather than other commercial game fish.

Baird's commission had no authority to set rules. Instead it worked with the states, advising them on fishing matters.[45] Nevertheless, Baird's position on conservation was clear. In his report for the year 1878, Baird wrote, "In the early days of the Republic, the entire Atlantic Shore of the United States abounded in fish of all kinds. Where cod, mackerel and other species are now found in moderate quantities, they occurred in incredible masses."[46] Powerful interests like the wholesale dealers at Fulton Market blocked federal fish conservation legislation during this era. Breeding replaced legislation since nobody opposed this strategy and there was apparently no downside. Politicians didn't have to pick winners and losers. By 1883, the commission was experimenting with the breeding of twenty-nine different species of fish or shellfish. This proved to be an excellent way to avoid confronting the difficult political questions raised by declining fish populations, since both sportsmen and fish wholesalers supported greater supplies of all food fishes.[47]

Fish dealers from Fulton Market dominated the New York State Fish Commission, which worked closely with Baird. Eugene Blackford played an important role in fish hatching efforts at both the state and federal levels. In his role as State Fish Commissioner, Blackford was the driving force behind the building of the state hatchery at Cold Spring Harbor. The breeding programs based there included salmon, smelts, brown trout, rainbow trout, and brook trout. These fishes and fish eggs were released in waters all across New York State.[48] The New York commission, with the full cooperation of the state legislature, was a model for other states interested in restocking depleted fish populations. Over time, the goal of hatcheries changed from restoring depleted fishes to making sure that there was enough game fish in any body of water for sportsmen to capture.[49] However, during the late nineteenth century, New York City's wholesale fish dealers still counted on the government to help make sure that they never ran out of popular fish.

The best example of this is shad, a popular fish with both sportsmen and dealers. As with so many other fish, technological improvements in the way they were caught threatened the overall supply. In this instance, those improvements involved the size and nature of the nets. In 1880, the naturalist John Burroughs wrote, "The ordinary gill or drift net used for shad fishing in the Hudson is from a half to three quarters of a mile long, and thirty feet wide, contains from fifty to sixty pounds of fine linen twine, and it is a labor of many months to knit one. . . . It is practically invisible to the shad in the obscure river current." In the narrow parts of the river, these nets could stretch almost entirely across.[50] No wonder the number of shad returning up the Hudson dropped precipitously before 1880.

With the assistance of its federal counterpart, the New York State Fish Commission successfully bred shad starting around this time, planting the shad eggs in the Hudson River in order to restock it. At the same time, the United States Commission of Fish and Fisheries brought in both eggs and fry from other rivers, like the Susquehanna, for the same purpose. In 1898, 13,927,730 were caught in the Hudson. That was almost nine million more than were caught in 1880 when the program began.[51] These efforts revived a fishery that had almost disappeared by the 1870s.

"There is no doubt that had the Fish Commission not undertaken the artificial propagation of shad it would have been exterminated long ago," explained the *New York Sun* in 1899.[52] This was literally true, as it appeared that the natural propagation of shad throughout the Eastern Seaboard had dropped to insignificant levels.[53] In the long term, shad hatcheries were not able to keep up with demand. By the post–World War II years, shad stocks fell into a serious decline even though artificial propagation continued.[54]

Although unpopular in some circles, conservation wasn't completely abandoned by politicians. Conservation laws failed at the federal level around the turn of the century, but there was a constant tension between sportsman and commercial fishermen in New York waters that led to some limitations on fishing for conservation purposes well into the twentieth century. Most of those restrictions were local because even at the state level, there were powerful interests working against conservation—including the wholesale dealers at the Fulton Fish Market.[55] An 1871 law passed by the New York State Legislature was only local in impact, banning purse seining in the part of Long Island where small-scale fishing thrived, but not elsewhere in the state.[56] Limiting nets in local waters or restricting the catch of some fish to particular seasons was a common small-scale solution, implemented in some places and not others. New York State passed low limits on the catch of menhaden, because the trawlers that caught them were a threat to sport fishermen and the public hated the smell that emanated from shoreline factories that processed that fish.[57]

The wholesalers at the Fulton Fish Market resisted any restrictions for conservation purposes on the catch they could sell. In 1893, twenty-nine of them sent a letter to the magazine *Forest and Stream* stating that prices hadn't increased at all in recent years.[58] Of course, wholesalers were just as capable of making money by charging more for scarce fish as by charging less for a plentiful product. Under certain circumstances, scarcity had the potential to make their business even more profitable.

When those dealers wrote that letter, much of the fish that went through the market left the city.[59] In the years following World War I, more and more fish that went through the market was consumed locally.

Demand for fish in the city increased steadily over the course of the twentieth century and persisted. In 1964, New Yorkers ate thirty pounds of seafood per capita, which was about three times that of the average American at the time.[60] However, that fish came from farther away than ever before. In 1925, the Associated Press reported that even as the market's fish were coming from far afield in recent years, "The market sends this tremendous poundage forth again over a much shorter radius." Most of it "scarcely gets a hundred miles from the East River."[61] A 1940 estimate suggested that only 10 percent of the fish that New Yorkers ate came from local waters.[62] Dealers had to source from farther away to keep supplies of particular fish steady or growing. This was a huge change from the nineteenth century, when most of the fish at Fulton Market came from relatively close by and was shipped all over the country.

Even though the fish that left the market had shorter distances to travel, new refrigeration techniques were required to keep it fresh before it got to the market. They were successful enough that consumers began to demand fresh fish at a reasonable price. "Stale fish should be avoided," wrote Sarah Tyson Rorer in her famous cookbook from 1902, "those kept in cold storage as well as those frozen in blocks of ice."[63] While volume dropped in the market as a result of the decentralization of seafood provisioning chains, dealers who sold the fish that consumers were willing to pay more for still made good money. That explains why a market with close to monopoly control over the nation's largest local population of seafood eaters didn't have to be particularly efficient in order to make money for the dealers who ran it. If the goods they provided cost enough, like green turtle or terrapin, they might not even have to refrigerate it at all.

7

TURTLE AND TERRAPIN

In 1889, the *Brooklyn Daily Eagle* asked Eugene Blackford where to find the best terrapin. Diamondback terrapins once ranged from Cape Cod to Texas and were a common sight at Fulton Market before 1890, by which time they had been hunted close to extinction because of the quality of their meat. Blackford's answer was Massachusetts because, as he explained, the farther south the terrapin came from, the tougher its meat was likely to be. A possible solution to the shortage was terrapin farming, but Blackford downplayed this possibility. "During the summer months," he explained, "there is no demand for terrapin and they can be purchased at a very low price. A few enterprising individuals buy them up at this time and put them in pens where the tide ebbs and flows. These terrapins seldom bring more than half price because of their toughness, fishy flavor and being destitute of eggs."[1] Blackford knew terrapins because by that point in time he had been selling them at Fulton Market for almost twenty years. Although terrapins are reptiles, not fish, they had been part of the regular offerings at the Fulton Fish Market for decades by that time.

Those terrapins had been caught in the wild. A couple of decades later, in parallel to the efforts to breed fish, a Savannah, Georgia, terrapin supplier named Alexander Barbee claimed to have gotten terrapins to breed in captivity for the first time. While that might have been true, it didn't change the fact that it took at least five years for a diamondback terrapin to be able to breed—maybe even ten.[2] Before diamondback terrapins became rare, anyone could eat them. Thanks to facilities like Barbee's, terrapin stew remained expensive but accessible for many decades. Despite the cost, it was an iconic Southern dish (assuming you include Maryland in the South) and popular wherever it was available, especially in New York City. In 1908, the *New York Times* claimed that 80 percent of the terrapins sold in the United States were consumed in the city. The paper also noted that the leading terrapin dealer in the country at that time had sold only fish just fifteen years earlier. Even though he doubted it was possible to breed diamondback terrapin in captivity, he had no worries about the supply.[3]

Many people use the terms "turtle" and "terrapin" interchangeably, but "turtle" should refer to turtles that spend their lives mostly at sea, while terrapins can be found on land or in swamps. Gastronomically, distinguishing the two is very important because each had a distinct market, even if they were sold in the same place. Diamondback terrapins are not large creatures, their maximum size being about nine inches in length. Green turtle—far larger, saltwater turtles picked up by sailors in the West Indies or off the coast of Florida—generally were cut into steaks or made into soup. Turtle soup first became popular in England, then spread to the United States. Of course, it was possible to make either dish with either kind of turtle (or even other kinds of turtles) and that sometimes happened, but these were the uses that first gave impetus to the two important and distinct American dishes. Turtle and terrapin provisioning chains stretched farther, earlier than those for other products sold at Fulton Market. By making money through selling more expensive items at a higher price, these two chains foretold a general move by the wholesalers located there toward quality over volume later in the market's history.

Today, green turtles are at the center of a worldwide conservation movement, but they were once fairly common in American waters. Civil War hero William Tecumseh Sherman, who had been stationed on the east coast of Florida in 1840, wrote, "They are so cheap and common that the soldiers regarded it as an imposition to be compelled to eat green turtle steaks instead of poor Florida beef or the usual mess-pork. I do not recall in my whole experience a spot on earth where fish, oysters and green turtle so abound as Fort Pierce, Florida." As late as 1884, it was still possible to catch a hundred green turtles in a single day off of Cape Hatteras in North Carolina.[4] Although green turtle was never a common food for most Americans, for much of the nineteenth century it was not rare. Getting it to the urban areas where people wanted to eat it was the problem that dealers in turtle had to solve. Terrapin and green turtle never made up a significant percentage of the total amount of product sold at Fulton Market. However, because they were often kept in fish cars—or in the case of particularly large green turtles, tied up out on a pier—they invariably attracted a lot of attention.

Green turtle first appeared on American tables as early as the 1700s but did not become a common meal component anywhere until decades afterward. It had to be served in restaurants and taverns because ordinary cooks did not have the means to deal with an animal that large and difficult to dispatch. "The subscriber will slice and carve one of the finest and fattest Green Turtle ever killed in New York on Friday next, 23d instant, at 6 o'clock A. M., at fish stand No. 24 Fulton Market," read an advertisement in the *Evening Post* in 1830. "It will be served in parcels to suit purchasers at one shilling the pound."[5] The next year, a different restaurant promised "two most splendid Green Turtle . . . in calipash, calipee [the upper and lower meat of the turtle], steaks and soups of a delicious flavor, with all the delicacies and substantials the market affords."[6] The need for advertisements reflects the relative rarity of the dish. "Green Turtles," explained the *New York World* in 1860, "are those objects which it is the ambition of every restaurant in Fulton street to have lying on their back before the door, and labeled, 'To be served upon this day.'"[7] They were bound tightly because live turtles would bite anyone with whom they came in contact, taking off fingers and causing other serious injuries.

As with terrapin, consumers paid good money for green turtle because it was considered very tasty. "It is not necessary to acquire a taste for turtle meat," wrote the *New York Times* in 1907. "To eat it once is to like its novel flavor."[8] Named for its green hue, the meat actually comes in three colors, which were generally referred to as "beef," "veal," and "lamb," the taste of which each one supposedly resembled.[9] Green turtle began regularly arriving in Fulton Market during the 1850s. By 1860, 60,000 pounds in a single year was a typical amount of turtle meat to arrive in New York.[10] Before 1860, the turtles came through middlemen. Then turtle fishermen in Florida began direct communications with dealers at Fulton Market in order to cut out the middlemen. Although Florida turtles appeared on tables in cities throughout the Northeast, New York had an exclusive deal to get all those caught by the only firm hunting them in the waters surrounding Key West, a particularly easy place to find green turtles.[11]

As was the case with many kinds of fish, green turtle had to be sourced farther away from New York City as time passed, in large part because wholesalers needed more to meet growing demand. By 1883, the city received between 150,000 and 180,000 pounds of green turtle meat each year. They came live for the same reason that lobster had to travel live—because if killed, the turtle would start to rot immediately and would spoil long before it arrived at the market. The added complication was that turtles almost always came in the summer, because during any other season, the water from the East River in their pens would have been cold enough to kill them.[12] Green turtles could range anywhere from 15 to 350 pounds. Some dealers would fatten up the smaller ones, feeding them mostly on watermelon rinds, before eventually selling them for soup.[13] Live turtles were delivered on sailing vessels, and steamships often brought them directly from Florida or Central America. They would then be sent out to other urban markets throughout the North, like Washington, D.C.[14] In Philadelphia, African American caterers with cultural and economic connections to the British Caribbean played an important role in establishing the market for green turtle.[15]

Unlike most fish, turtle weren't processed at Fulton Market until canning facilities opened there during the early twentieth century. Slaughtering the turtle was usually the responsibility of whomever cooked the soup. This could be a very onerous responsibility. "The day you intend to dress the turtle cut off its head"; read *Mrs. Rorer's Philadelphia Cook Book* from 1886, "and to do this properly you should hang up the victim

FIGURE 7.1 "Green Turtle Soup Today," by P. Frenzeny. Green turtles had to be killed by chefs right before cooking because their meat spoiled quickly. The cruel treatment these large, live turtles received helped spur a campaign against the dish, but dealers in Fulton Market sold these animals for as long as a market for them persisted. *Harper's Weekly*, January 5, 1884.

Author's collection

with its head downwards, use a very sharp knife and make the incision as close to the head as possible. You must not be surprised at seeing, many hours after the decollation, the creature exhibit extraordinary signs of muscular motion, by the flapping of his fins."[16] The expressiveness of the turtle made this difficult task even more difficult. "I saw a great turtle lying in a restaurant the other day," explained a reporter from the *New York Press* in 1906, "flat up on his back, his head pillowed on a cigar box and his flippers tied with stout strings. He was alive, of course, and yet eyed with a look of sullen but puzzled defiance the group which stood about him while the proprietor of the place explained, illustrating with touches of his foot, the way to which the creature was to be presently cut up and the different manner to which the various parts would be cooked."[17] Eventually, dried turtle meat and canned turtle meat became popular in some circles, but these shortcuts did not offer quite the same level of prestige as bringing a giant animal into your kitchen and treating your guests with not just food but an experience.[18]

Outside of New York City, green turtle was substantially more rare, making it even more expensive than it might have been otherwise. Turtle soup recipes appeared in some of the better cookbooks of this era, but the price of turtle meat as well as the difficulty faced by anyone trying to kill a turtle themselves meant that real turtle soup was usually restaurant food. The green turtle soup at Nash and Crook's Park Row restaurant included two "knuckles" of ham, a "knuckle" of veal, two carrots, two turnips, six onions, two dozen cloves, two dozen allspice berries, six blades of mace, and a dozen bay leaves—and that was just the stock. Each time it was ordered, the cook had to fry onions, add flour for thickening, then add red pepper, lemon juice, sherry, brandy, and Worcestershire sauce and caramel for color.[19] Nash and Crook must have felt confident publishing this recipe because they understood that very few people would ever attempt to make it at home.

Another way to get green turtle was at a catered banquet. The most famous were the dinners of the Hoboken Turtle Club. That organization was founded in Hoboken, New Jersey, in 1796 by a former captain in George Washington's army and survived into the 1930s.[20] During its heyday in the 1880s, the club had relocated to New York and bought its turtles in Fulton Market. The businessmen, politicians, and other civic leaders ate turtle steaks for breakfast and turtle soup for dinner, along with countless other dishes whenever they gathered for their well-reported "meetings." In 1886, *Frank Leslie's Illustrated Newspaper* reported that the group kept the shell of the first turtle it had eaten, back in 1796, to mark what it considered to be that auspicious occasion, but by then, the success of turtle distribution through the Fulton Market had made turtle soup much less of a rarity than it had been.[21]

Between 1880 and 1890, green turtle fishing grew sharply, especially in the Gulf of Mexico, from 24,000 pounds in 1880 to almost 500,000 pounds in 1890. Now there were schooners devoted entirely to the turtle trade. The first green turtle soup canning facility appeared (ironically) in

Fulton, Texas, in 1869. Ten years later, that facility became a factory and in 1881 greatly expanded. Canned turtle soup was a way for Americans with social aspirations to eat the same way that the rich did. The green turtles that weren't processed in Fulton were sent to Galveston and then on to Fulton Fish Market. By 1895, the Texas and Florida green turtle fisheries had been completely depleted.[22]

Being much smaller and more numerous, terrapin started as a common dish and became more expensive over time as supplies dwindled. While terrapin stew was a country dish, like green turtle, it had roots in African American culture. Food once fed to slaves, it ascended through African American cooks like Emiline Jones, who started life in slavery as a house servant but eventually cooked terrapin stew for Presidents Garfield, Arthur, and Cleveland.[23] Nonetheless, this dish might never have moved beyond New Orleans and Baltimore if not for the diamondback terrapin trade in the Fulton Fish Market. Both green turtle and diamondback terrapins had once been cheap, but each became expensive as their meat became more popular. By the 1890s, both were rare delicacies in the United States. Fulton Fish Market supplied them both well into the twentieth century.

There was no commercial terrapin industry before 1845. As terrapin stew became popular in northeastern cities, the first terrapin hunters began to catch them around the Chesapeake Bay area. Many of these suppliers were rural African Americans with hunting skills who sought terrapins for extra income. Also in 1845, a North Carolina man perfected the first terrapin net. A North Carolina lighthouse keeper quickly used that invention to catch 2,150 turtles at the same time. Other terrapin hunters used dogs to track larger turtles, but that kind of hunting was frowned upon because the dogs tended to destroy the turtle's eggs.[24] Terrapins were divided into three grades: counts, heifers, and bulls. Counts had shells that were over six inches in length. Heifers had shells between five and six inches long. Bulls had shells five inches long or smaller.[25] You could make turtle soup out of terrapin, but most went for stew. Only female diamondbacks were considered a delicacy, so only they fetched the highest prices. Males did not reach the largest sizes, and their meat was considered inferior.[26]

A six-inch terrapin could make six quarts of stew. However, actually making quality terrapin stew was supposedly a difficult task.[27] In 1896, an anonymous "epicure" gave his recipe to a reporter from the *New York Times*. It required dropping the terrapin in boiling water for twenty-two minutes. When cutting the meat from the shell, the gall bladder had to be removed or it would ruin the rest of the meat, but the liver was chopped up and included in the mix. The sauce included flour, milk, butter, salt, cayenne pepper, and cinnamon. Sherry or madeira went directly onto the terrapin meat before the sauce and the meat were mixed into the stew. Notably, this epicure told the paper that they had received this recipe from "an old Virginia mammy," which suggests the huge difference in the racial and class origins between green turtle soup and diamondback terrapin stew.[28] Only when both green turtles and diamondbacks became rare did they both become luxury foods.

The ingredients in the sauces may have been to the key to recipes for both kinds of turtle. In reference to terrapin stew, the legendary naturalist William Temple Hornaday wrote in 1922, "In flavor I think it has been greatly overrated . . . With all the good things that go into a terrapin stew, and champagne for sauce at three-fifty a bottle, almost any animal would taste good."[29] A similar point was made about green turtle soup in 1878, when a member of the Hoboken Turtle Club with access to the kitchen admitted to a reporter from the *Times*, "This is turtle soup of the best kind, but there's not much turtle in it. It wouldn't do, you know. Too much turtle spoils turtle soup. . . . It would be so rich, nobody could eat a cupful of it."[30] This may explain why many observers, despite singing its praises, invariably noted that turtle was an acquired taste.

Unlike with green turtle, New York was not the chief market for terrapin during the late nineteenth century. Philadelphia and Baltimore probably bought more, and that is reflected by the menu at Fulton Market.[31] There were three different kinds of terrapin stew available: Philadelphia, Maryland, and Southern. One recipe for Southern stew (albeit not cooked at Fulton Market) included hard-boiled egg, country butter, thick cream, flour, salt, nutmeg, and "a glass and a half of amontillado." The Maryland style available at Fulton Fish Market included dry sherry and thick cream and cost $10 a quart. The Philadelphia-style stew probably contained snapping turtle, which for some reason was popular there despite the risk that snapping turtles posed to chefs' fingers.[32] Many of the terrapins caught during the nineteenth century came from the Chesapeake Bay and stayed in Baltimore. Just take a look at the mascot of the University of Maryland. That meant that there were fewer discerning gourmands in New York who could tell the difference between diamondback terrapin and other turtles. This explains why it was pretty easy to substitute other turtle meat for terrapin if the provider was so inclined.[33]

By the 1890s, the dish that once fed slaves had become primarily associated with millionaires. "Time was when there was plenty of terrapin in all the bays and estuaries of the ocean," noted Eugene Blackford in 1895. "Greed led fishermen to take up the young, and now a terrapin is almost worth his weight in gold."[34] Between 1886 and 1896, there was a leap of 100 percent in the cost of terrapin from the North and 50 percent for those sourced south of the Chesapeake. Hotels and restaurants could only sell them as special orders. Even though they were much smaller, terrapins, like green turtles, took a long time to grow to maturity—about ten years. Keeping them that long was a difficult business proposition, and it proved very difficult to get them to breed in captivity.[35] That alone explains why their price remained high well into the twentieth century.

Terrapin farms were supposed to meet the large demand. These turtles, unlike sea turtles, were small enough that farming them was at least a theoretical possibility. Unfortunately, supply could never keep up. After the Depression, the price was too high to maintain the demand that terrapin farms needed to survive. Terrapin stew remained on some menus, but even after it became rare, diamondback terrapin numbers never rebounded.[36] The same could be said for demand for the terrapin. "They are dying off," one dealer told Dr. Robert Coker of the Bureau of Fisheries in 1920, before the advent of Prohibition. "It is the old buyers only that come to me for diamond-backs."[37] Joseph Mitchell visited the

Barbee terrapin farm in 1939, then followed the supply line back to Fulton Market.[38] That indicates that terrapin stew survived both Prohibition and the Depression, but it was never again quite as popular as it had been during the late nineteenth century.

Unlike fish, or even terrapin, green turtles were such large, expressive creatures that their treatment at the Fulton Fish Market inspired a considerable backlash among animal rights advocates. Henry Bergh founded the American Society for the Prevention of Cruelty to Animals (ASPCA) in 1866. Shortly afterward, the state legislature passed a law that protected animals from cruelty. Such efforts quickly spread around the country. The New York law allowed the ASPCA to appoint its own agents to enforce the law and arrest perpetrators. Bergh made himself a badge-wearing officer against animal cruelty. Fond of bringing publicity to his movement, he often let newspaper reporters follow him around. He attracted a lot of influential enemies as he attracted attention, most notably the showman and circus impresario P. T. Barnum.[39]

In 1866, Bergh boarded a schooner named *Active* and arrested its captain for cruelty toward the green turtles it was delivering to Fulton Market. The cargo was flipped upside down on their shells and bound with ropes, and because those ropes had pierced their flippers, had visible wounds after weeks on the ship.[40] In 1871, Bergh, acting as an agent of the state, brought charges against Eugene Blackford and one of his drivers for cruelty to animals because Blackford's wagon was carrying "a large-sized turtle laying on its back; its head was jammed against the side of the wagon, the blood was oozing from its belly, the back was crushed in and the bottom of the wagon was stained with gore. The turtle was alive at the time, but it seemed to be dying."[41] In 1876, Bergh brought similar charges against Blackford again.[42]

Bergh lost all these cases. Nobody ever went to jail for cruelty against green turtles. The defense in these disputes invariably noted that carrying the turtles on their backs was both standard practice in the trade and the least cruel way to transport live turtles because their shells were harder than their bodies on the other side. There were also long debates over whether turtles could actually feel pain. In response to the constant pressure from Bergh and the ASPCA, Blackford (both as a shipper and as part of the New York Fish Commission) modified industrywide practices to make the green turtle industry seem less cruel. At Bergh's suggestion, Blackford made each of the turtles he kept a "pillow," a salt bag filled with sawdust. Dealers also experimented with other ways to secure the turtles' flippers that didn't do so much damage. Apart from these modifications, Bergh argued that his attracting attention to the plight of green turtles was a watershed moment in the cause of animal rights. Keeping these turtles this way may have been legal, but it was obvious to anyone who saw them that it was not morally right.[43]

In fact, the whole turtle provisioning process was cruel from start to finish. Because green turtles were so big, they were hard to handle. Because they had to be shipped live to stay fresh, they were treated cruelly. For example, they had to have their mouths tied or they'd bite the people trying to handle them. Such treatment may not have been that much more cruel than that of any other animal slaughtered for food, but their huge size guaranteed that green turtles would attract far more attention than smaller creatures. In 1885, a reporter for the *New York Times* visited the Bahamas in order to explain how green turtles were caught for the New York City market. "We have slits cut in the ends of her fins and her fins tied tightly together, so that she cannot make any fuss . . . In less than a quarter of an hour, our captive is secure and we start in search of another victim." The smaller ones who were not turned over in this way on the beach were speared in shallow waters, and a hole was made in their shell for the spear to hold them in place for the entire long trip to the city.[44]

That trip was often deadly. In 1903, a reporter interviewed a man at Fulton Fish Market with the nickname Turtle Bill. Big and burly, Bill had handled turtles of every sort at the market for the last fifteen years, which implies that he worked for a turtle dealer rather than being a dealer himself. "Most of the animals are landed here in lots of twenties or thirties from the coast schooners that touch turtle countries," Bill explained, referring to green sea turtles. "On the whole we figure on two shipments of turtles happening into port each week." Although that may seem like a lot of turtles, many didn't make into New York alive. "The animals are brought here and stored to find out whether they are alive or not," Bill explained. "By tapping the shell of the turtle, he'll bob out his head to see what's the trouble. But most of the poor brutes die on the way up from their countries."[45]

Green turtle was expensive enough that suppliers could make money even if a lot of their cargo died en route. However, there were unique problems associated with keeping a tropical reptile alive in New York City, even in the summer. "This climate and its waters do not tend to prolong the life of the turtle," wrote the *New York Times* in 1895, "and after being placed in the tank it does not live more than five or six weeks."[46] Of course, if the turtle was sold to a restaurant or hotel before that point, it didn't even live that long. During the early history of the market, live fish never inspired this kind of sympathy. Neither did dead fish, which were transported to and sold in the market during almost all of its history. Green turtles attracted more attention than any kind of fish that went through the market, even though the turtle business was never more than a small part of the overall operations.

By 1940, most of the green turtles sold at the Fulton Market ended up in Chinatown.[47] The same became true for terrapins, at least for a while. "I imagine you're under the impression that millionaires buy most of our terrapin," remarked the manager of a Fulton Market turtle firm to Joseph Mitchell in 1939. "If so, you're mistaken. The terrapin business was hard hit by prohibition, and it has never gotten on its feet again, and for years the poor Chinese laundryman has been the backbone of our trade." They ate it as an aphrodisiac. "Better than monkey glands," claimed Alexander Barbee. Another firm simply claimed that eating terrapin promoted a longer life. "Mr. [Francesco] Castellli believes that turtle and terrapin meat is the most healthful in the world," wrote Mitchell of the manager of Moore and Company, which handled both.[48] Moore and Company formed during the 1850s and remained one of two major turtle retailers in the Fulton Fish Market into the mid-twentieth century. Castelli dealt in both terrapin stew and turtle soup made from green sea turtle.

Earlier dealers, like Eugene Blackford, had kept their turtles tied down outside in the market. Castelli ran a warehouse at 147 Beekman Street that kept turtles and terrapins and used them to make canned soup.[49] In 1947, when a reporter for *LIFE* magazine interviewed Castelli, he told them that his firm "rendered 5,000 Caribbean turtles into 6,000 quarts of broth annually." Turtle soup had not only survived but also, at a price of one dollar a quart at many food stores, become more accessible than ever.[50] Besides explaining the state of the turtle industry at this point in time, the fact that the turtle and terrapin business had moved from the market into the neighborhood around the market indicates the path that a lot of other fish-related businesses were taking and would continue on later in the market's history.

The trade ended when green turtle was certified as an endangered species under the Convention on International Trade in Endangered Species of Wild Fauna and Flora, adopted in 1973.[51] Terrapin stew remains on American menus because restaurants have substituted non-endangered turtles for the diamondback terrapins of the nineteenth and early twentieth centuries.[52] Perhaps it's a good thing that the market for green turtles has disappeared because of the effect that they seem to have had on the people whose businesses depended upon them. As *LIFE* reported, "Now more and more soup turtles have to be kept alive in a dim, dreary storage room, where they eat nothing and emit loud, shuttering sighs. This grieves . . . Castelli, a tenderhearted man. 'Sometimes I'm in my office late at night,' he says, 'and I hear those turtles sighing in the next room and I don't know. I just don't know.' "[53] As was the case during the nineteenth century, turtles got a lot more sympathy than fish even after they were moved out of sight.

Although green turtles and terrapins were not fish in the technical sense, the U.S. government considered both part of fisheries with good reason. Green turtles were hunted from boats, then placed in pens after being apprehended. Terrapins could be caught with dredges and nets in their swampier habitats and also were placed in pens.[54] Like fish in the days before ice, both had to be kept alive for much of their transit because their meat spoiled quickly after they died. Most importantly, the environmental effects of catching both required management by the federal government. With fewer turtles and terrapins available in the first place, the regulation of this fishery became an important issue far faster than with any fish species. Conservation took hold better than it ever has with many popular fish, and the basic principle behind conserving turtles and terrapins is the same as with any other endangered species. Turtles and terrapins illustrated the pitfalls of unfettered fishing, but seafood wholesalers never heeded the warning.

Because turtles and terrapin reproduce much more slowly than fish do, the effect of the trade in both had a greater impact on the species as a whole than most fishing did on any fishery. While the dealers certainly treated green turtle in a cruel manner, the long-term tragedy arising from the market for these specialties was that they were hunted to the brink of extinction.[55] Like with so many fish species, when available supplies began to run out, wholesalers at Fulton Market continued to source them from new places farther away. Only government intervention prevented both turtles and terrapins from disappearing entirely. From the perspective of the turtle dealers, this behavior is easy to understand. If the shortage of turtles or terrapins increased prices enough, paying others to hunt them almost to extinction created the possibility of charging even more. The unbending focus on profit explains why all the wholesale dealers continued to make money even as fish populations shrank drastically. The success of this strategy for staying afloat financially helped them ward off the trouble caused by technological developments up and down the provisioning chain.

8

FREEZING, COLD STORAGE, AND IMPROVEMENTS IN TRANSPORTATION

Frank W. Wilkisson was a wholesale dealer in the Fulton Fish Market who had been a friend of Al Smith's back in his market days.[1] In 1939, Frank left a nice fresh fish outside his stall in the market with a sign above it that read, "I should be worth 27 cents. But the public does not want to buy me. I could make a fine dinner for some one. Oh my! What will become of me?" The next day, the same fish remained, but Wilkisson had placed a new sign over it: "Yesterday I was worth 27 cents. Today only 12 cents just because the public wouldn't buy me yesterday." The next morning, a new sign over the same fish read, "I have no value as food now. No, not even a penny. Maybe some fertilizer company will take me. And two days ago I was worth 27 cents."[2] Fruit and vegetables are perishable foods, but their value doesn't decrease that quickly. Meat can actually improve when aged under the right conditions. No class of food deteriorates in value as fast as fish does.

The famous motivational speaking trainer Dale Carnegie used Frank Wilkisson's story to illustrate what he considered to be an important principle of business. In his column, he explained to newspaper readers around the country that Wilkisson wanted to get the fishmongers in the market to work together to solve the common problem of their product quickly losing value. Instead of Wilkisson telling them they must work together, the fish and the signs supposedly gave them the idea to organize themselves. The result was the Fishery Council, a group that issued regular press releases designed to convince consumers to eat more fish so that their product would sell faster.[3] In retrospect, a better solution might have been to improve the cold storage infrastructure around the Fulton Fish Market so that nice fresh fish wouldn't lose its value quite so quickly. The failure of the wholesalers to build enough cold storage or freezing facilities throughout the twentieth century may have been the defining trait of the whole market.

This problem took time to emerge because it took decades for ice refrigeration to be replaced by a modern cold storage system, and ice alone was enough to preserve and supply the city with a huge variety and volume of fish. Noting the extraordinary variety of sea creatures at Fulton Market was a common trope in stories about the place. Even if those fish were dead, their availability in New York was noteworthy for the average newspaper reader. By the mid-twentieth century, fishermen, processors, and wholesalers throughout the country used various refrigeration technologies to effectively preserve seafood of all kinds. This development greatly increased the competition for consumers outside the New York market and forced the dealers in the Fulton Fish Market to focus on local buyers, altering the daily life of the market forever.

Better refrigeration and cold storage anywhere along the fish provisioning chain meant less waste. Less waste meant more supply, which usually translated into lower prices. The availability of ice meant more fish could be sourced from farther away, which increased supply and also meant lower prices. Ice could also help fish dealers defy the seasons. Many species only appeared at Fulton Market at particular times when they could be caught, and then had to be eaten quickly. With ice, rudimentary cold storage was possible for the first time. When consumers came to realize that the fish they ate would taste all right no matter what time of year they ate it, demand for fish increased. Although cold storage and freezing have some negative effect on taste, the better these technologies became, the harder it was for consumers to tell whether they had been used. However, New York City never embraced these technologies the same way that newer hubs in the fish provisioning system did.

When it comes to the freshness of fish, temperature matters as much as time. Higher temperatures encourage the bacteria that make the fish decay. The longer the fish are dead, the more opportunity there is for that temperature to rise—at least if the fish are being refrigerated entirely by ice.[4] The increased use of ice meant a decrease in live fish being landed at the Fulton Fish Market by boat, since catches that landed elsewhere could be protected on their journey to New York City. Obviously, fish were still coming out of the water, but the easier it was to arrest their decay on the boat and in transport, the less urgent it became to get them to the next stop on the supply chain. Ice helped, but could only do so much.

Freezing fish—and keeping them frozen—made it far easier to keep them indefinitely. However, the original Fulton Fish Market was never particularly interested in selling much frozen fish. This created an opening for wholesale fish dealers elsewhere to serve customers, especially cost-conscious customers, everywhere. Frozen fish could keep indefinitely, which meant that it could travel farther. Because the Fulton Fish Market lagged behind other fish markets around the country in implementing freezing technology, the volume of fish going through the market eventually declined as the technology improved. This continued the long decline of the Fulton Fish Market as the central location for fish provisioning in the United States.

The first step in getting from refrigeration with ice to effectively frozen fish was the huge increase in ice production made possible by technological improvements in every aspect of that business during the late nineteenth century. However, the situation in New York City was different than that in the rest of the country at that time. The city was dominated by the "Ice Trust" (the Knickerbocker Ice Company, later the American Ice Company), which had almost exclusive control of the ice harvested from the Hudson River. As a result, there was little machine-made ice produced in the city even as ice machines became more effective during the 1890s.[5] That's why wholesale fish dealers depended upon third parties for their ice, and those needs were considerable. In 1894, a dealer told a reporter for the *Evening World* that the market used a "low average" of 25,000 tons of ice each year.[6] Some went for storage purposes.

FIGURE 8.1 "Fulton Street Market Ice Chopper," by Sol Libsohn, 1938. Both the Fulton Fish Market and the fishermen who serviced it needed ice to keep their fish fresh. Because it arrived in large blocks, they needed an ice chopper to cut it into smaller pieces.

Museum of the City of New York

Some went for shipping fish far outside the city. That number didn't include all the ice used on fishing boats that didn't unload at Fulton Market or on shipments of seafood originally landed in places like Boston or Baltimore that got sent to Fulton Market for sale.

The dealers in Fulton Market also depended upon ice for display purposes. Keeping the skin moist with water that gradually came off exposed ice helped keep up appearances.[7] If nothing else, the presence of ice at least cut down on the smell that fish emitted since it stayed fresh longer. A lot of ice was essential for the first effective methods of freezing fish. Freezing them in the open air during cold weather had been a common preservation practice that went back centuries. Enoch Piper of Camden, Maine, got the first patent for an invention that froze fish in 1862. He laid out the fish on racks, then laid pans of crushed ice and salt on them, then let them sit for twenty-four hours. Those fish were then placed in another room, insulated by tubes with a mixture of ice and salt running through them. Ice and salt could freeze fish because together they lower the temperature of anything below freezing. Piper used his invention to build the first cold storage room for fish in the United States, near Fulton Market on Beekman Street. Over time, a series of improvements were made, and the size of cold storage rooms gradually increased as a result.[8]

A fish chilled with ice might look a little different than freshly killed fish, but it usually remained recognizable as a fish. A frozen fish is either surrounded by ice or at the very least frozen stiff. In many instances, the fish was already cut up into a fillet so that it resembled other kinds of meat, at least superficially. In short, a frozen fish may be recognizable as having once been a fish, but it has noticeably changed since entering the stream of commerce. It was possible to use ice to freeze a whole, intact fish, but that was not nearly as effective as modern freezing techniques would become. This explains why consumers often resisted frozen fish in its early days.[9] The first cold storage equipment using mechanical refrigeration that was exclusively for fish wasn't built until 1892 in Sandusky, Ohio. This was the future of refrigeration technology, but it wouldn't be perfected until after the turn of the twentieth century. Until then, frozen meant deeply chilled with ice rather than frozen solid in the modern sense, but that was still enough to vastly expand the market's range both coming in and going out.[10]

A cold chain is a term coined by refrigerating engineers for the infrastructure that keeps perishable food fresh from its point of production to its point of consumption.[11] Filling in a complete cold chain took an enormous amount of innovation because one kind of refrigeration might work in stationary warehouses but not in rail cars. At the same time, different kinds of seafood had different ideal temperatures, and storing all of it properly took a lot of experimentation. Improvements in the refrigeration, freezing, and cold storage of fish began outside of Fulton Market but changed the way business got done there by either creating competition or eventually forcing the wholesalers who ran the place to adapt to the new technologies. Unlike improvements in the efficiency of fishing, these changes weren't particularly welcome by the wholesalers.

Rapid freezing was necessary for the chain to be completely effective at the point where fish came out of the water. Clarence Birdseye, of frozen food fame, worked on the technology of fish freezing during the mid-1910s.[12] However, others were more effective at it earlier. Like so many inventions, this technology was evolutionary rather than revolutionary. The Fulton Fish Market was receiving frozen fish that had been caught off Labrador as early as 1907.[13] By 1918, the head of the Bureau of Chemistry at the U.S. Department of Agriculture could write, "It has been thoroughly demonstrated that freshly caught fish, when properly frozen and maintained at a temperature below freezing, will keep for months without changing in flavor or appearance."[14] The problem was freezing the fish immediately after it left the water and keeping it that way through the entire cold chain, all the way from the boat to the dining table.

Accomplishing that required further improvements in the technology of flash freezing. The purpose of flash freezing is to capture and maintain the taste of fresh fish from the moment that it comes out of the water. Unlike earlier experiments with freezing fish, this was being done on boats even before Birdseye began his experiments on land. In 1897, the *Fishing Gazette* reported on a ship called the *Tillid*, a vessel with a freezer in its hold. "The fish, alive and kicking, as it entered the scupper," the *Gazette* reported, "is frozen stiff, and a very few minutes longer suffices to coat it with a thick frost, in which condition it remains until taken out and sold." More importantly, the *Gazette* wrote, "It will be observed that the idea of freezing fish during the summer months in a vessel specifically fitted for the purpose is distinctly new."[15] The Bureau of Fisheries imported America's first quick freezing machine for fish from Europe in 1918. That machine inspired further innovations throughout the industry.[16]

By the mid-1920s there were multiple ways to freeze fish. By the mid-1950s, freezing mostly occurred at onshore processing facilities.[17] Most of that fish was cheap, highly processed fillets sold through grocery stores rather than through fish markets.[18] In general, freezing fish greatly contributed to the process of making it much more like any other commodity, in that it could be stored indefinitely and transported anywhere without serious degrading.[19] Industrial scientists recognized that the more rapidly freezing could be accomplished, the more effective it would be. There were still many important advancements in the technology of fish freezing during the rest of the twentieth century. The problem that still had to be solved was that regular freezing inhibited bacteria, but it also created ice. When that ice crushed nearby cell structures, the texture of the fish's flesh becomes mushy.[20] The solution would be flash freezing. Even after it became possible to freeze fish, keeping it frozen on its journey from boat to plate required numerous variations of freezing technology that worked on boats, trucks, and other irregular, mobile spaces.

In order for wholesale dealers to even out the supply of fish and meet an irregular demand, they had to store fish as well as freeze them. While ice allowed fish to travel safely across space (assuming there was enough available throughout its route), mechanical refrigeration became a way for fish to travel through time by making it possible to preserve fish out of season. Early cold storage methods only worked in particular rooms with lots of equipment. 1920s-era innovations like refrigerated trucks and very rudimentary refrigerated rail cars (which required gas burners and silicon gel to keep their cargo cold) completed the cold chain needed to keep fish fresh through its entire journey.[21] All these developments together allowed wholesalers to stabilize the price of some fish by releasing them onto the market in such a way that the price wouldn't drop precipitously when there was an oversupply. Of course, operating cold storage units cost money, so fish could not be stored indefinitely. Nevertheless, improvements in this technology went a long way toward regularizing the irregular rhythms of nature.

In the early days of cold storage at Fulton Market, there was a fine line between iced and frozen fish. By the mid-1870s, Eugene Blackford had his own ice cellars to keep valuable fish fresh when the price had unexpectedly dropped.[22] The advantage of this strategy, despite what cold storage did to the flesh of the fish, was that it helped even out demand. A Connecticut paper wrote in 1887: "During these seasons of plenty the dealers resort to the process of freezing, and manage to have a supply on hand when the market is depleted." The main problem with this strategy was the cost. "The process entailed great expense and constant care and watchfulness," the paper continued, "but enabled many dealers to supply different kinds of fresh fish to their customers, when these particular varieties would otherwise have been out of the market."[23] Whereas the cost of cold storage using ice added up quickly, mechanical refrigeration improved the quality of freezing and cost less for the dealer.

That changeover happened slowly in New York because the city was used to cheap ice cut from the Hudson River. The Brooklyn Bridge Freezing and Cold Storage Company built the first mechanical refrigerating plant near Fulton Market in 1886, but since this technology still wasn't completely effective, it served as more of an experiment than a turning point in the history of the market.[24] Because it had the majority of the capacity around Fulton Market—as had been the case with the ice companies before it—the firm set its rates very high. Any fish that went to these facilities had to be moved by truck. It didn't help that fish couldn't be stored with other kinds of perishable food because the smell tended to seep into everything over time. Even then, fish was a difficult product to preserve well because it is so easy for it to taste off when even the slightest mistake is made in the way it is handled.[25]

Gradual improvements in cold storage between approximately 1903 and 1909 made this industry an effective way to store all perishable foods, including fish.[26] However, the shortage of cold storage capacity around Fulton Market was a persistent problem. None of the main market buildings had any cold storage capacity before 1920. Fish dealers complained about the lack of cold storage facilities in the city throughout World War I. When met with unexpected surpluses, they had to truck the fish to the outer boroughs or New Jersey (or farther) to have any hope of saving it. Then they had to send it back to Fulton Market so that it could reach its final destination.[27] Some Fulton Market dealers had to keep their frozen product at facilities in Buffalo, New York, or Newport, Rhode Island.[28] Obviously, this led directly to a lot of waste. The extra costs associated with the inconvenience of storing fish and of storing it so far away meant that wholesalers seldom used this technology.

Cold storage capacity in the United States grew rapidly after World War I, but not at Fulton Market.[29] "The large bulk of sea food at Fulton Market never gets into a freezer," explained a reporter for *Power Boating* magazine in 1923, "as it is simply packed in cracked ice, sold and consumed before getting into cold storage."[30] Fulton Market fell farther behind the times as the Bureau of Fisheries fiercely promoted frozen fish to American consumers during and after the war. Frozen fish had been widely seen as an inferior product, but fishing trade journals noted that demand increased in the cities where the Bureau of Fisheries targeted its marketing campaign.[31] The Fulton Fish Market continued to be the central link in the fish provisioning chain for the greater New York area throughout the rest of the century, but it would never duplicate the influence it had had over the nation's fish consumption habits during the late nineteenth century because its poor technological infrastructure hindered its growth, while other markets around the country took advantage of these improvements.

The quality and popularity of frozen fish are among the best explanations for the decline of volume at the Fulton Fish Market over the course of the twentieth century. Ice was an adequate tool for fish preservation for the entire nineteenth century, but by two decades into the twentieth century the needs of the market had changed. Modern freezing techniques and improved cold storage capacity for fish made it possible for wholesalers to even out supply and demand. People wanted more fish at different times, often for religious reasons, like Lent or just on Fridays. When the technology improved enough that frozen fish tasted as good as fresh to most people, fish wholesalers could supply uneven demands at reasonable prices without a lot of waste. Fresh fish could be a lucrative business for wholesalers, but only frozen fish could feed the masses.[32] However, without enough proper freezing and cold storage facilities, it was difficult to produce around the Fulton Fish Market. As a result, new provisioning chains developed that bypassed the market entirely.

Improvements in transportation outside the market had some of the same positive effects as improvements in freezing and cold storage for fish going through it and helped make up for the deficiencies at the Fulton Fish Market, to at least some extent. Better transportation limited the time that fish had to decay, which helped make up for the fact that there were only limited cold storage facilities available in and around the market. Better transportation was not visible in the market (where boxes were moved by hand trucks well into the late twentieth century), but its impact was. By cutting the time between when a fish was landed and when it was eventually consumed, railroads prevented spoilage and improved the quality. Refrigerated trucks later did the same. The ability to reach more markets faster also made it easier for wholesalers to sell the surplus of fish that trawling had made possible. When airplanes came into the picture, just the ability to reach more markets at all helped wholesalers elsewhere. These kinds of changes enabled the Fulton Fish Market remain profitable, even as its lagging adoption of modern refrigeration slowly made the facility obsolete.

The changes in transportation began in earnest with the spread and increased speed of railroads during the late nineteenth century. By 1890, a New York dealer could telegraph Gloucester at 8 p.m. and have a shipment delivered via a nightly Boston-to-New York train the next morning. It was always in perfect condition thanks to advances in ice refrigeration technology. New York dealers could even arrange shipments from Boston to Chicago in the same manner because their role in the fish provisioning chain was that central.[33] After the turn of the twentieth century, there was another regular train carrying cod and other fish from Fulton Market up to Boston.[34] Another train carried fresh halibut from Vancouver to Fulton Market. That trip took seven days.[35] The market required halibut and salmon from the other side of North America because of the shortages of both fish in more local waters.[36] In 1904, menhaden steamers that had been landing in Connecticut, repacking their fish, and sending them to Fulton Market by boat began landing on Long Island. From there, the catch was sent via the Long Island Railroad straight to the city.[37]

By 1899, only eight or ten vessels a year still brought their catch into Fulton Market alive.[38] As early as 1909, 70 percent of the product at Fulton Fish Market came via rail cars refrigerated with ice.[39] Because there was no direct rail connection at Fulton Market, those fish had to be transferred to Lower Manhattan by truck. By the 1920s, the docks on the East River had fallen into decline and Fulton Market was essentially cut off from the water, far earlier than it might have been if not for developments in refrigeration.[40] After fish mostly stopped arriving by boat, there ceased to be any compelling reason to keep a fish market there at all, other than the fact that moving it would have entailed a lot of trouble for all the parties that bought and sold goods there.

In fact, Fulton Market was not particularly well suited for receiving anything by rail. The fish in cars from the north were unloaded in Harlem and sent down to the market by barge. Other refrigerated fish off rail cars came to the market from other stations (including New Jersey) by cart, or later, truck. All this handling drove up prices and decreased quality.[41] In turn, much of that same fish left town via rail car, incurring the same transportation costs in the other direction.[42] It would have been impossible for fish to arrive in New York from farther and farther away without improvements in refrigeration and transportation along those journeys. By 1925, 85 percent of the fish came by rail. Between the early 1930s and early 1950s, there was another rapid shift from rail to refrigerated truck.[43] The extension of these supply lines from one end of the country to the other provided incentive for catching yet more fish.

However, the same improvements made the Fulton Fish Market less important to America's fish industry in general. As a 1911 history of New England fisheries explained, "Improvements in the refrigerator car service made it possible for the fresh fish of New England to be delivered a thousand miles from Gloucester. But the same service brought the fresh salmon of the Columbia and the fresh halibut of Alaska in open competition, in Boston and New York markets, with Penobscot River salmon and with halibut fresh from the banks."[44] An advertisement for a Philadelphia fish wholesaler in 1918 put it succinctly: "We are in a position through our connections with the Fisheries throughout the country to execute any order, regardless of size—a single package or car lots."[45] As a result of these changes, some of the larger wholesalers in the Fulton Market began to branch out to other cities. Others began to specialize in new kinds of seafood. Some simply went out of business, but the market as a whole continued operating.

New communications technologies accelerated this gradual decentralization. As early as the 1930s, fishermen could telephone fish markets in ports up and down the East Coast as soon as they landed in order to figure out where they could get the best price for their catch. Obviously, fish that was presold this way stayed out of the open marketplace that came together most early mornings, so the significance of this technological innovation was difficult for most reporters to appreciate. The fishermen could find better places to ship their catch, bypassing the Fulton Fish Market entirely.[46] This decentralization at both ends of the provisioning chain threatened the wholesalers' near-monopoly control over the New York market.

The similarities and differences between the supply chains for beef and fish after the turn of the twentieth century demonstrate difficulties that food provisioners had when adapting to change. The beef supply chain shipped live animals to Chicago, where they were killed, dressed, and sent east in rail cars refrigerated by ice. Fish were shipped alive to Fulton Market, where they were processed and sent out around New York and around the country on ice. The food provisioning system for fish arrived earlier then meat (which wasn't perfected until the 1880s), and like the system that went through Chicago, it had two important components: a long trip to the central hub and, at least in the case of some fish early on, another long trip after. In Chicago, everything that could be done with a cow's carcass was a source of money for the packers. In New York, the wholesale dealers often sold a wide range of fish and other sea creatures to expand their business. And like in Chicago, the provisioning system began to decentralize after the turn of the twentieth century. In the case of beef, urban development in places closer to the steers made it easier to process beef on the plains where ranchers fattened it. The difference between beef and fish was that large meatpacking firms could move their businesses to earlier parts of the supply chain, shortening the total distance and saving time and travel in the process. Some of the seafood wholesalers in New York City invested in operations elsewhere, but they were never large enough to pick up shop and move entirely.

None of this decentralization would have been possible without effective freezing. Freezing gave the parties up and down the provisioning chain the time they needed to find the best outlet for whatever fish they acquired. Previously, seafood of all kinds had gone through Manhattan because that was where the most buyers and sellers congregated. This accelerated the speed with which seafood moved on the market and limited the amount of time it had to decay over the course of the provisioning chain. Effective freezing took time out of that equation. By not improving its refrigeration and freezing facilities enough, the Fulton Fish Market left its competitors in other cities a huge opening to expand their businesses at the expense of New York dealers.

After World War II, improvements in refrigeration technology brought refrigerators and freezers to retail fish stores and supermarkets. Refrigerated trucks served these stores, thereby completing one end of the cold chain.[47] On the other end, improvements in both transportation and refrigeration technology merged into a single development in 1953. The first factory trawler was a British vessel named the *Fairtry*. At 240 feet long, it was the largest fishing boat in the world at that time. With a crew of between seventy and eighty sailors, it could stay out for twenty days at a time. Much of that labor went toward processing and freezing the fish on board. Although this was more expensive than doing the processing on land, it eliminated the cost of dealing with the bycatch because only the targeted fish got frozen and sent ashore.[48]

The technology to freeze fish at sea began with the tuna fleet during the 1930s. By the mid-1950s, trawlers could no longer operate profitably near the coasts of the United States because previous boats had severely depleted the supply of fish. Therefore, trawlers had to be able to freeze fish because they had to be able to stay out to sea longer.[49] To make onboard freezing economically viable under these circumstances, the *Fairtry* had to be roughly twice the size of other trawlers.[50] Machines below deck processed the fish it caught before it returned to dry land.[51]

The problem for any fishing boat that depended on ice to refrigerate its catch was having to carry the ice. Even though much of it would melt, all that ice left less space for fish. Freezing fish on board not only saved more space for the catch, it preserved the fish better than ice because rapid freezing limited the ice crystals that formed on the product.[52]

The *Fairtry* was followed by similar trawlers owned by the Soviets and Germans, who concentrated on catching cod in the North Atlantic.[53] Thanks to new freezing methods and boat designs, these trawlers and those that immediately followed them could fish in places where they had never been before because they could preserve the fish over long distances. This included the Atlantic coast of the United States, just outside of American territorial waters.[54] IT was all part of a worldwide increase in the intensity of fishing in the years following World War II. The number of fishing vessels throughout the world grew from 451,000 in 1970 to 885,000 in 1995. By 1965, the Soviet Union alone had 106 factory trawlers, many of which operated at the edge of American territorial waters.[55] These trawlers formed the basis of alternative supply lines that could put cheap fish on people's dinner plates without going through the Fulton Fish Market.

The effects of all the improvements in fishing and fish provisioning during the 1950s and 1960s upon fish stocks were immense. The hake, the herring, the mackerel, and the whiting were just a few of the fisheries that these gigantic vessels devastated.[56] The problem with trawlers wasn't just the catch; it was the bycatch. Gigantic nets scraping the ocean floor took every fish in their path, and all those unwanted fish were invariably discarded and usually died. Improved freezing techniques had made all that slaughter economically feasible. Otherwise, none of the desired fish could have been marketed in their home countries. American seafood companies contributed to this depletion by starting joint ventures with foreign firms so that they would always have access to fish, even if it came from far away.[57] These larger firms that became dependent upon imported fish operated at scale, not by selling their catch at stands in Lower Manhattan.

Refrigeration and freezing are closely related, evolutionary technologies. They change gradually, with contributions made by inventions from around the world, rather than in sudden bursts. An improvement might make them more effective or cheaper, or might make them possible in new places, like a rail car or a fishing boat. At the end of World War I, this technology had turned fish from a resource into a commodity. Further improvements in the refrigeration and freezing of fish would make it possible to maintain the fish in a good state from one end of the provisioning chain all the way to the other. This created a global marketplace that had never existed before. Contrary to its reputation, the Fulton Fish Market adapted to this new environment. However, had the dealers there been a little less conservative in outlook, it might have adapted far more effectively.

9

FROM THE BROOKLYN BRIDGE TO THE FDR DRIVE

Between the early and late nineteenth centuries, the Fulton Market neighborhood changed from an area dominated by the port into an increasingly diversified commercial area. A letter writer to the *Brooklyn Daily Eagle* remembered in 1912 that before the Brooklyn Bridge was built, "It was a busy locality, the stores on Fulton street from Pearl street to the river being occupied by large retailers and jobbers of groceries, wines, foreign and domestic fruits, willow and wooden wares, coffees, teas and various household commodities. On Saturdays they remained open to the midnight hour, because many passengers, by stages from the theaters or other belated ones stopped to make purchases."[1] The businesses on Schermerhorn Row reflected this fact. Although its first occupants were overwhelmingly merchants, by the mid-nineteenth century the row included grocers, barbers, eating houses, and bars.[2] In 1882, Thomas Edison opened the first central station in the United States in a building at the corner of Fulton and Pearl Streets. These industrial facilities, like the fish market, made the neighborhood a predominately industrial area.

Although few people lived around Fulton Market from the 1840s going forward, the neighborhood was not unpopulated. People from Brooklyn, small towns surrounding the city, and especially sailors (who stayed there, but their presence was transitory) came to enjoy the many vices available. Taverns and brothels served the sailors and the immigrants who lived in the many five- to seven-story tenements in and around the neighborhood.[3] 279 Water Street contained a bar called The Hole in the Wall, where the female bouncer had a tendency to bite the earlobes off of patrons who misbehaved. Supposedly they kept a jar of pickled earlobes at the bar.[4] During the 1860s, the founder of the gang the Dead Rabbits, Kit Burns, set up Sportsman Hall in the neighborhood. Thousands visited there over the years to bet on animal fights.[5] Despite the competition from the other side of Manhattan, the neighborhood remained a working port into the 1930s. A lot of the buildings housed fish-related businesses like cooperage companies, fishery supply companies, and transportation firms.[6] The people on the streets included longshoremen, stevedores, and draymen. Tattoo artists and dancing girls served the needs of the men who worked the ships that still docked in the area.[7]

Then the whole orientation of the metropolis began to change when the Brooklyn Bridge opened in 1883. The bridge changed everything in this neighborhood because the Manhattan ingress ramp was both north and many blocks west of the fish market. People came to prefer taking the bridge to the ferry (and the Brooklyn Bridge has a footpath so that people can cross for free). This meant that a neighborhood they once had to pass through became inconvenient. As Joseph Mitchell explained, "On account of the ferry, Fulton Street was like a funnel; damned near everything headed for Brooklyn went through it. It was full of foot traffic and horse-drawn traffic day and night, and South and Fulton was one of the most ideal saloon corners in the city."[8] When the infrastructure that dumped people into the neighborhood surrounding the fish market changed, the neighborhood changed with it. Indeed, it became an industrial novelty just as such neighborhoods were becoming nearly extinct in Manhattan. Only during the late twentieth century did it start to look like the rest of the island again—a transformation that came directly at the fish market's expense.

Later bridges like the Williamsburg Bridge (built in 1903) and the Manhattan Bridge (built in 1909) made it even easier to bypass the neighborhood when going between Brooklyn and Manhattan. The

Pennsylvania Railroad's tunnels under the East River, built around the same time, only aggravated this trend. In 1899, a reporter for the *Brooklyn Citizen* visited Fulton Market to see how it had changed over the years since he had last been there. "The rush and roar of the thronging thousands as they passed by to and from the ferry was gone," he wrote, because the Manhattan side of the bridge deposited pedestrians many blocks west of Fulton Market. "There was no glut at the river gates. The great bridge had spun its airy web overhead and caught travel... while in the meantime the ferry had grown a legend."[9] The Fulton Ferry line closed in 1924, and when it did, the *Brooklyn Times Union* noted, "Not one Brooklynite in ten thousand has used the old Fulton Ferry in several years."[10] Lack of foot traffic destroyed the last vestiges of the retail business in the market.

So did the tendency of the whole population of New York City to move uptown throughout the nineteenth century. The enactment of the famous grid plan in 1811 illustrated this trend. Even as the Port of New York thrived, the center of population of the city moved farther away from what had once been its gateway along the East River. By 1864, half of the population of New York City lived north of Fourteenth Street.[11] In 1919, the real estate developer Harry Hall wrote of downtown New York City:

> A student of business movements will see that in the last quarter of a century a constant and irresistible pressure... has steadily forced an expansion of building development in every direction until the entire face of the district bounded by Fulton, Beaver, Pearl and Church streets has been changed beyond recognition. Step by step, and street by street the old order has given way to the new; the landmarks of a previous generation have succumbed to the greater efficiency of the modern architect and engineer, and within this movement have come increasing land values, with consequent profit for the fortunate real estate owner and operator.

The Fulton Fish Market was just east of that district. Hall argued that the progress that affected the rest of downtown would inevitably cross

Pearl Street and enter the neighborhood around the market, but that did not begin for another forty years.¹²

The same dynamics that cleared out the neighborhood around Fulton Market affected the mix of stalls inside. The consumers who had once shopped at the retail market had all moved away. Practically the only customers left were buying fish they intended to sell elsewhere rather than consume themselves. The wholesale fishmongers had become more important over the course of the nineteenth century, but there were still some butchers and fruit sellers in Fulton Market at the time the Brooklyn Bridge opened, and a few of them persisted. In 1902, when a delegation from the City of Boston's Committee of Markets visited, that had changed. Almost 50 percent of the stalls in the market were empty, "it being impossible to obtain tenants for them." Of the stands that were rented, a large percentage were occupied by just four wholesale fish dealers. Many of the vacant stalls were in good locations for whatever foot traffic remained in the area.¹³

The abandonment of all other food sales except for wholesale seafood around this time marked the biggest transformation in the market's long history. "Time was when this was the busiest retail market in the country," remarked a retail stall owner to a writer from the *Outlook* in 1911. "Do you remember how, besides the fish-stalls, this market was crowded with small restaurants, and all of them jammed with customers? And there was one big restaurant, too, where people came from all over the country to get their fish suppers. The owners spent twenty-five thousand dollars, and it was sure a swell place. But they've gone uptown, long ago." When that same writer interviewed the proprietor of a booth that sold a purplish seaweed called Irish pulse, he was told, "[My father] made money in this place; but I don't know why I've stayed here. Just because I grew up in the market, I suppose. It's dead as a door-nail now."¹⁴ Because of this change, wholesale fish businesses got new buildings from the city and had room to expand their operations in underutilized buildings throughout the neighborhood. In the long term, however, this situation set the stage for the market's eventual relocation.

As the neighborhood changed, the market had to change with it. When those visiting Bostonians asked the remaining dealers what had happened, the response was: "This change they attribute to two causes—first, that people had either moved some distance away from the market, or do not have occasion to be in that vicinity; and, second, that they could get their wants supplied, more conveniently by the various grocery and provision stores, practically none of which were in existence twenty or twenty-five years ago."[15] Changes in both the way people moved through New York and the way food was distributed throughout the city worked to turn Fulton Market into the Fulton Fish Market. Although the business of selling fish wholesale continued to grow for some time afterward, the role of this market as an institution in New York life never really returned.

Even before the turn of the twentieth century, individual firms took over buildings in the neighborhood to use as offices and sometimes as bases to do additional business outside their market stalls. In the 1894 letter to *Forest and Stream* in which twenty-nine dealers denied that fish species were being depleted, eleven of the signatories listed their addresses as being on Beekman or South Street, instead of Fulton Fish Market.[16] The New England Fish Company expanded its operations to New York City, bought a building in the neighborhood to use as office space, and helped with the shortage of cold storage in the area by building its own modern facilities.[17] Inter City Fish built its own building on Water Street in 1914.[18] Oyster dealers started moving into buildings in the neighborhood near the river after the wholesale oyster market moved there in 1912.[19] After 1925, many fish wholesalers and processors occupied spots on Schermerhorn Row, converting former hotels and boarding houses into storehouses with cement floors and refrigerators.[20] On the mornings when the market was open, operators of these establishments often did business in the streets. So did the firms that had stalls in the market itself. When the market wasn't operating, the dilapidated buildings that the seafood dealers occupied appeared essentially empty. Apart from the smell, who wanted to live in a neighborhood that attracted the most people in the middle of the night?

The decline of the neighborhood around Fulton Market was also part of another broader urban trend involving the decline of the entire New York waterfront. The neighborhood, like the entire east side of Manhattan, was hurt when ocean liners and other large ships began to dock on the Hudson River side of Manhattan because the water there was deeper. This was obvious as soon as 1902, when all the freight yards on the island were along the Hudson. Fulton Market was serviced almost entirely by trucks in this era because there was no other way to get fish when the boats stopped coming, and there was no way for trains to stop anywhere near there. Without much traffic, the city let the quality of the now-inadequate piers significantly degrade.[21] By 1950, two-thirds of the ships entering the harbor docked along the Hudson River, not the East River.[22]

FIGURE 9.1 "Fulton Fish Market [variant]," by Berenice Abbott, 1936. By the late 1930s, the last Fulton Market retail building had been mostly taken over by wholesale seafood firms.

Museum of the City of New York

As the retail business at Fulton Market degraded, so did the condition of what had once been its main building. After a fire in 1911, the *Times* wrote of Eugene Blackford's 1883 replacement for the original Fulton Fish Market building, "The exterior of the market presents a dilapidated appearance. There is scarcely a whole window of glass at any point on its four sides . . . The woodwork is rotting, the roof broken, and there is an appearance of general decay everywhere about the building." In response to the changes around the neighborhood, the city abandoned the building entirely in 1914.[23] When a syndicate of fish and oyster dealers bought it in 1917 to use as office space, the *Real Estate Record and Builders' Guide* called it "little more than a shell."[24] Nevertheless, there were still stalls for wholesale fish dealers in the structure after the sale.[25]

While it took decades for retail to rebound (at what would then be known as the South Street Seaport), the wholesale fish business thrived. In 1907, the wholesalers built a new building at the foot of Fulton Street. It would come to be known as the "Tin Building" even though it was built with galvanized steel, and it was built along the same design as the 1869 building.[26] Its spot between Piers 17 and 18, on top of a platform over the water, made it possible to deliver fish directly by boat. It had three floors: stalls on the first floor, dressing rooms and offices on the second floor, and storage on the third. There were no doors on the South Street side, just a shelter for unloading wagons, and later trucks.[27]

The new building and its structure were signs of the transformation of the business in the market from the retail to the wholesale trade. "In the years that have elapsed since 1869 [when the Fishmongers' Association formed]," explained the *New York Tribune* in 1903, "the fish trade has grown beyond all expectations. More merchants have gone into business, and, unable to procure room in the Fulton Fish Market, they opened stalls ashore until the whole block bounded by Beekman, South and Front sts. and Peck Slip became devoted to the selling of fish."[28] However, the wholesale fish business did not require nearly as many stands as the retail business did, making the open parts of such buildings far less important than they once had been. That explains why a firm as big as Blackford's built its own building in the neighborhood around this same time.[29] Indeed, by the turn of the twentieth century, a lot of fish market business spilled out into the neighborhood during its early morning hours of operation.

In 1909, the dealers who hadn't been able to get space in the new building formed two new associations, the New York Wholesale Fish Dealers' Association and the Independent Wholesale Fish Dealers' Association.[30] The former moved into a new building located between the docks of the Fulton Ferry and those of the Hatford Steamships line. The latter built their own building adjoining it and also occupied in 1909. The dealers there hadn't been able to get into either of the other two buildings.[31]

Dealers in freshwater fish established their own market near the end of Peck Slip, in a series of old brick buildings that dated back to when Peck Slip had been a working pier. Some of them had been warehouses. Others were places where shipowners lived. When the area surrounding the slip was filled in, it became a street for boat-oriented businesses. Then it turned over to scrap metal dealers, rag sellers, and the like. Freshwater fish dealers started moving into the empty buildings around 1899. Their best customers were Jews interested in fishes such as haddock, perch, and carp.[32] This was the only part of Fulton Fish Market that actually auctioned fish. It went into steep decline after the 1920s mostly because consumers began buying their gefilte fish from supermarkets. That way, they didn't have to process the whole fish themselves.[33] That market lasted until 1963, when Consolidated Edison bought several of the market buildings and erected a substation there.[34]

Since the windows on so many of the upper floors of the buildings surrounding the market were boarded up, the whole neighborhood looked abandoned during the day—and it mostly was. "On an errand to one of the offices in the sprawling fish market," wrote a *New York Times* reporter in 1938, "you are likely to find your way to an ancient unwashed building with swarms of cats and kittens running around the ground floor among barrels of fish. Top a rickety flight of dark wooden stairs you wander along a cold corridor to a little front room barely furnished."[35] The photographer Walker Evans, who shot the remaining warehouses built near the water during the nineteenth century, wrote of these mostly abandoned structures, "Under sunset illumination, these gaunt exhausted boxes, dotted with their iron shutters of old blue or splotched

FIGURE 9.2 The 1909 Fulton Fish Market building, built for a small coalition of dealers, slid into the river in 1937 after its pilings were hit by a fishing boat.

South Street Seaport Museum, New York, NY, gift of Vincent J. Tatick, H. Carter, Inc.

russet, are a delight to lovers of paradigmatic Americana. Their very hardware has style and patina—great wrought-iron hasps, fanciful metal stars that are the cross-support termini, ponderous door fittings that would suit a prison."[36] Foreswearing the larger market complexes, newer seafood and seafood-related companies worked at night when fresh fish was available or did business on the sidewalks outside their buildings during market hours. That's why it could be difficult to tell that the neighborhood had gone commercial.

The Seaman's Church Institute of New York, which housed sailors on shore leave, dominated the neighborhood long after most of the residents left. Thousands of seamen were temporary residents. They patronized the stores, wandered around the docks, and took naps in local parks. Ship chandleries and cordage stores persisted even as most of the boat traffic moved to the Hudson River side of Manhattan and the fish market began to depend almost exclusively on trucks because many non-fishing boats still used the docks. During winter, a sizable community of people actually lived on barges docked near the market. The cargo those barges transported the rest of the year couldn't travel in water clogged by ice, so whole families made the piers their home. These scenes had defined the area for decades by the late 1930s, but it was still obvious that the neighborhood was changing.[37]

The decline of the neighborhood around Fulton Fish Market came even though New Yorkers were eating more fish. By the late 1930s, the market itself had become a place that only served its own local market. Although fish still arrived from as far away as Florida, the buyers who flocked there came mostly from retail stores and markets within a 200-mile radius.[38] Only 20 percent of the seafood going through the Fulton Market was shipped to other locations.[39] The market was still important—especially to New Yorkers—but it was no longer the center of nearly every fish provisioning chain in the United States. Nevertheless, selling more fish to people who lived nearby proved very lucrative for many wholesalers. Moreover, thanks to modern communications, they could still deal in fish that never actually went through the market itself. That explains how they prospered even as the neighborhood declined.

Much of that prosperity, however, was skimmed off by organized crime. Joseph "Socks" Lanza got a job at the Fulton Fish Market when he was fifteen. He was arrested for juvenile delinquency at the age of seventeen. Three years later he was arrested for burglary, which he pled down to unlawful entry. After that, in 1922, he helped organize a United Seafood Workers' Union local at the Fulton Fish Market.[40] For most of its history, it would be known as Local 359. By 1925, half of the laborers in the market were Italian Americans, and Lanza became their business agent.[41] He was closely connected to the Genovese crime family.

The power of the union derived from its monopoly control of loading and unloading fish at the market. Because fish spoils easily and there was very little access to cold storage around the market, the failure to load or unload any fish could completely ruin a business. This was true for both the wholesalers who received fish to sell to retailers and the retailers who needed that fish loaded into their trucks. The threat of violence against both groups if they wouldn't pay to have the union members load or unload contributed to the effectiveness of this lever. Starting in the 1920s, fish dealers also paid the mob for insurance against the theft of fish by the men who handled it.[42]

The exact origin of the mafia involvement in the fish market is obscure. Either Lanza helped bring Italians into the market or he saw that they were getting jobs there anyway and took advantage of that fact.[43] No matter how the mob arrived, its existence at the Fulton Fish Market first emerged into public view during the investigation of Judge Samuel Seabury into the activities of Manhattan District Attorney Thomas C.T. Crain in 1931. The Seabury Committee aired testimony related to payments from wholesalers to the union in order to protect them against theft by union members loading and unloading their fish. Retail merchants also testified that they had to pay the union to watch their trucks so that they wouldn't be vandalized, presumably by the same union members.[44] Crain investigated the presence of racketeering at the market but brought very few charges because "Socks" Lanza had supposedly physically threatened witnesses.[45] Others recognized that the wholesalers willingly paid Lanza's union, which made bringing charges difficult.[46]

That same year, a grand jury made two different sets of sweeping charges against actors in the fish market. One set of charges involved racketeering and extortion, forcing retailers to buy freshwater fish at inflated prices and forcing fishermen to sell at low prices through threats of violence.[47] In 1936, a U.S. attorney's investigation came to an agreement with many of the most respectable saltwater fish dealers in the market. The mob-connected union was demanding both a $10 per boat landing fee to dock at the market and fees from trucks bringing fish into the market from elsewhere.[48] The fees were Lanza's charges that supposedly went to the union's "benevolent fund."[49] This was ultimately devastating to the long-term future of the market because other seaports in other cities with no landing fees began to get more business.[50] The wholesalers, like the truckers and the peddlers from whom Lanza extracted fees, were technically the victims of extortion (even if they willing accepted Lanza's services), not its perpetrators.

The dealers were, however, fixing prices. Unlike the racketeering, which was only revealed after the Seabury investigation, the price setting by the dealers was out in the open. "Although the commercial fisherman on Long Island may have some idea as to the general market price of his fish," explained a New York State Conservation Report from the 1930s, "he has no accurate knowledge of the total value of his shipment until he actually receives his payment from the assigned dealer in Fulton Market. The prices of the fish change daily and almost hourly, and are so dependent upon quality and quantity factors that no stabilized price can be established." To make matters worse, the assignment of the catch to the dealer who sold it was done by lottery, so the wholesaler could rip off his supplier one day and still be assured more fish the next.[51] More commonly, the fish dealers would simply average what they paid to fishermen in commissions over the course of the day so that everyone got the same price, even though fish sold later in the day often went for less because of quality and sometimes concerns about supply.[52] This is why the wholesalers had to stop selling fish on consignment in 1936.

The press conflated the price setting and the racketeering. The price fixing was a civil violation of antitrust laws rather than a criminal offense, but the papers made it hard to tell the difference. "Those Fulton Fish Market firms occupying the true Fulton Fish Market are not racketeers," explained the *Elizabeth City* [NC] *Independent* in 1935, "and North Carolina fishermen can continue to consign their shipments to them with every assurance that they are doing business as usual and that there is no change in the status of their established reputation for fair and honorable dealings. But the press reports looked bad to those who don't know the facts."[53] Lanza was initially convicted in 1935 for racketeering only in the freshwater fish trade. He served two and a half years, but that sentence was extended six months when he pled guilty to racketeering in the trade for other seafood. In 1943, Lanza was convicted of labor extortion and sentenced to between seven and a half and fifteen years in prison.[54] Because the Genovese crime family had a limitless supply of enforcers, mafia control of the market continued without him.[55]

In 1936, the north end of the 1909 building, built for the group of less important wholesale dealers, fell into the East River. Had it happened during the time of the night or day when people worked there, this might

FIGURE 9.3 Mayor Fiorello La Guardia poses with a 300-pound halibut he received at the opening of the last new Fulton Fish Market building in 1939. *World Telegram & Sun* photo by C. M. Stieglitz.

Library of Congress, Washington, D.C.

have caused a considerable number of casualties.[56] The crash destroyed a dozen stands and put still more wholesalers temporarily out of business.[57] The immediate cause of the accident was that part of the foundation of the pier the building stood on gave way. The rumor around the market was that a trawler captain with a fondness for drink, exiting a nearby dock shortly before the accident, had collided with a wooden pole that held the building up. Whether this was true or not, people had been worried about the pilings before the collapse. One contractor working for the city had already concluded that the building was "past the stage where further reinforcement would help." The city quickly replaced that building with a new one, which opened in 1939. The wholesalers were so

FIGURE 9.4 Walker Evans, [Waterfront Brick Buildings in Fulton Market, New York City], 1933–34. The awnings near the bottom allowed the firms based there to conduct business in the street rather than in the traditional market buildings.

Image copyright © The Metropolitan Museum of Art. Image: Art Resource, NY happy with the new structure that they presented Mayor Fiorello La Guardia with a 300-pound halibut at the opening ceremony.[58]

The changes in the Fulton Fish Market were part of other modernization efforts in the neighborhood and adjoining neighborhoods. Most notably, north of the Brooklyn Bridge, the tenement buildings that had been the home of so many fish market employees like Al Smith were being replaced during the 1930s with new public housing. The city began constructing a new promenade with parks and playgrounds along the East River. Closer to the market, the planning began for a new East River Drive. Primarily intended to speed traffic up and down Manhattan, it was also intended to remove traffic from around the fish market to make it easier for trucks to go in and out of the area. This required widening South Street and building overhead ramps to speed the movement of fish from the piers into the market.[59]

The city finished what came to be called the FDR Drive in 1954, which saved the neighborhood from traffic but also left it increasingly isolated.[60] That road is elevated at South Street, just two feet above the top of the Tin Building.[61] This kept the Tin Building and much of the rest of the area in permanent shadow and divided what had been Fulton Market in the nineteenth century from the part of the Fulton Fish Market that still operated and was in public buildings. Ironically, the road helped protect workers in the market from the elements. "Thank God they built it to protect us from the weather," one market man later remembered.[62] It also protected the area from development for a time, since potential tall buildings with a view of a road that busy were not exactly prime real estate.[63]

Starting in the mid-1950s, the neighborhood surrounding the market became a haven for artists who could not afford studio space uptown. These included Jasper Johns, Robert Rauschenberg, and Cy Twombly. One of the qualities they liked about working there was the relative privacy afforded by the area's nonresidential status.[64] By 1960, it was said, you "could count the population on your hand."[65] That same year, the *Daily News* reported, "The area around the fish markets, an area of cold-water flats, now houses an estimated 400 writers, sculptors and artists. [Stanley] Cranston [who ran the Fulton Art Gallery] explained that the soaring rents in Greenwich Village sent the art colony eastward. Fulton St. runs right into the Wall St. district and the big insurance company buildings. It is nothing for executives to drop in at lunchtime, pick up a painting, and, if it isn't a special exhibit, cart it home after work."[66] That proximity to Wall Street helps explain the next wave of change to hit the neighborhood during the following decades,

The decline of the Fulton Ferry Hotel, one of many businesses on Schermerhorn Row, can represent the decline of the neighborhood in general. Located across from the entrance to the ferry, it once got passengers in transit both in the hotel and at the bar. After the Brooklyn Bridge took the ferry traffic away, the clientele became mostly drunks and old men on pensions.[67] Another hotel on Schermerhorn Row closed during the 1930s. The last of the neighborhood's boarding houses closed during the 1940s.[68] "The area is largely now occupied by commercial slums," explained David Rockefeller when he proposed putting the World Trade Center there in 1960. "I don't know of any other area in the city where there's as good an opportunity to expand inexpensively."[69] The abandonment of the neighborhood helped the market survive the gradual decentralization of the fish provisioning chain by helping wholesalers save on real estate costs. However, no part of Manhattan would stay essentially empty forever.

After 1960, sensing that the real estate under and around the fish market would be valuable if put to a different use, Wall Street and insurance companies began to move in on an area that had once been primarily industrial. Although Manhattan had once been a center for industry, the real estate got so valuable that much of it moved to the outskirts of town, if it stayed in town at all. The industries connected to New York as a port stayed much longer than others until moving off the island entirely, often all the way to New Jersey. After the port diminished in importance, the city turned its back on its waterfront. The Fulton Fish Market remained in place, but its location became more and more precarious as time passed. It is easy to see how the residents of an island could forget that their waterfront even existed when you realize how polluted the surrounding water became after the turn of the twentieth century. The clearest indicator of that pollution in terms of products sold at the market were the city's once-beloved oysters.

10

POLLUTION AND THE DECLINE OF NEW YORK'S OYSTER INDUSTRY

When various fisheries became depleted near the shores of New England, fishermen traveled farther away to find supplies for dealers like the ones in Fulton Market. The intensification of pollution in the waters around New York City caused the same response. As early as 1885, the water around Fulton Market itself was so tainted by sewage that the fish kept in the fish cars could only be left there for a few hours without spoiling their taste.[1] The situation was no better in the water surrounding what would soon be known as the outer boroughs. Oil and gas companies, especially John D. Rockefeller's Standard Oil, were notorious for dumping the by-products of the refining process directly into the waters surrounding the city. Household and industrial waste mixed together to create stinky messes of pollution in all the city's waterways.[2] There had once been an active sport fishing industry in New York Harbor. Even sharks appeared there regularly, but increased pollution was one of several factors that led to their disappearance around that time.[3]

Around the turn of the twentieth century, the population grew to such a level that the waste the city dumped into nearby waters significantly changed the ecosystem. This pollution, combined with the effects of overfishing, caused fish stocks to crash.[4] Human waste had always been a problem, but the ever larger population in New York made the problem much worse as the twentieth century passed. The water in the Hudson, for example, became so filthy that the city built an elaborate system to bring in drinking water from the Catskills. This required digging a tunnel under the Hudson River so that the fresh water wouldn't be contaminated by polluted local water.[5] The growth of manufacturing farther up the river made things even worse. After World War II, the General Motors plant near Tarrytown dumped so much paint in the river that you could tell what color they were painting the models coming off the assembly line that day by the color of the water in the river.[6]

Fish taken from polluted water generally taste bad. This proved particularly true for shad, which started to taste of oil at the turn of the twentieth century as the pollution around New York City worsened.[7] However, the clearest effect of pollution on the local seafood supply involved oysters. The effects of both human and industrial pollution on oyster beds were apparent as early as 1885. By 1902, those effects became impossible to ignore.[8] Once an oyster attaches to a surface, it become stationary. If stationary oysters happened to be located too close to a sewer opening, they probably became contaminated. Typhoid fever is a waterborne disease caused by a bacterium carried in human waste. The typhoid fever bacilli that entered oysters came in through whatever water the shellfish happened to be filtering. This made it harder to definitively pinpoint oysters as a cause of the disease because there were usually other possibilities—like water or milk (which was often watered down in this era). Documenting the oyster supply chain better proved the best way to limit the damage from any such outbreak.

The first typhoid fever outbreak in the United States directly tied to oysters occurred at Wesleyan University in Connecticut in 1894 and sickened thirty students.[9] The *New York Times* reported, "The results of this inquiry appear as complete and conclusive as those of certain notable investigations which have revealed the cause of typhoid epidemics in germs carried by milk from dairy farms where the disease had prevailed. The explanation is clearly along the lines established by sanitary science."[10] The first typhoid outbreak around New York City directly tied to oysters involved shellfish raised near a sewer in Jamaica Bay, south of Brooklyn, in 1904. In this case, the relationship between typhoid fever and oyster consumption was never really in doubt. The question was how far an oyster bed had to be from sewage so that the pollution would be diluted enough for the oysters to be safely consumed.[11]

Although European scientists made the rather obvious connection between typhoid fever and oysters during the 1880s, American public health officials resisted drawing this same conclusion for another two decades.[12] That's why a Fulton Market oyster salesman could argue in the pages of the *Brooklyn Times Union* in 1902, "It's all rank nonsense that there are typhoid fever germs in the oysters. Anybody who knows anything about how they grow knows that there is tidal action continually clearing the waters in which oysters are found, and even if you shovel a barrel of rubbish and stuff overboard onto an oyster bed, the tides would distribute it so widely that you wouldn't be able to ever get it together again."[13] Fulton Market was a central stop in that supply chain, but because the wholesalers there did not have the same store of knowledge that they had had about fish during the nineteenth century, they denied that the product they were selling put people at risk. Ironically, the nutrients in the sewage actually promoted oyster growth.[14] The Fulton Market dealers were obviously self-interested when it came to the science of oysters and disease, but their denials could only stave off disaster for so long.

The publicity surrounding the relationship between typhoid and pollution devastated the oyster business around New York City. In 1908, one dealer claimed that the industry had already lost $1,000,000 because of media coverage of this link and that sales around the country had already decreased by 25 percent.[15] In 1909, the city's Department of Health reported, "The inspection of the oyster supply in the City of New York is being rapidly systematized. Charts are being prepared of all the points along the coast from which oysters are obtained. In connection with these charts there is maintained a card index of the location from which all dealers procure their oysters and a cross index of the names of all dealers to whom oysters are shipped from any particular locality."[16] The dealers could only deny the problem for so long because their consumers began to vote with their wallets long before public health officials finally began to act.

The Bureau of Chemistry in the United States Department of Agriculture definitively proved the connection between typhoid and oysters in its investigation of an outbreak at a banquet in Goshen, New York, in October 1911. What changed was that health authorities were able to pinpoint oysters as the cause by examining particular banquets and noting that the attendees who ate the oysters got typhoid fever, while the people who didn't were fine.[17] The officials traced those particular oysters to a wholesaler in Fulton Market, then back to their supplier. They then traced oysters from the same wholesaler to a second, smaller outbreak in Suffern, New York, and then a third outbreak in Newburgh, New York. Those last oysters were bought directly from the supplier that sold the batch that started the first two outbreaks, not from the Fulton Market wholesaler.[18] However, the Bureau of Chemistry's report on the 1911–12 outbreaks did not appear until 1916, and it still took time for regulators to act definitively in response.

Jamaica Bay was the source of the water that infected the oysters in a number of these outbreaks. At the time of the 1911–12 events, Jamaica Bay produced between 500,000 and one million bushels of oysters a year, and twenty-nine public sewers discharged into the bay. To make matters worse, the underwater plots where oysters were grown were leased to growers by the state of New York, and the filthy water there had already been documented. The Bureau of Chemistry's Report concluded, "Considered as a whole, Jamaica Bay may be considered as one great basin into which many millions of gallons of human sewage find their way over a portion of the shellfish grounds through the medium of polluted water."[19] Some of the beds there were literally drowned in sewage sludge.[20] Despite this fairly obvious recommendation, the city Department of Health didn't ban the sale of oysters taken from Jamaica Bay until April 1916.[21]

The four-year delay between the outbreak and the banning came from an effort to differentiate the waters where oysters were floated from the waters in Jamaica Bay where oysters were actually harvested.[22] The floating of oysters was already somewhat controversial for being deceptive, and authorities took direct aim at the practice when the areas where it was done became polluted. In 1912, the Department of Health banned the floating of oysters only in the waters closest to human habitation because they thought that was where all the risk was. They also forced the wholesalers to learn whether and where the oysters they sold were floated from their suppliers.²³ New York City banned the floating of oysters altogether in 1919 because of the possibility that they would be polluted.²⁴

In 1904, the Fulton Market Fishmongers Association displaced a smaller oyster market that had existed at the foot of Beekman Street so that it could expand its oyster-related facilities.²⁵ In 1912, all the remaining barges were relocated to the East River at the foot of Pike Street, which became the city's official wholesale oyster and clam market.²⁶ The process of consolidating all the seafood-related industries of New York was complete. Yet, with the natural beds mostly depleted and pollution an obvious problem even before the 1911–12 typhoid outbreak, the oyster industry had to rely on increasingly ineffective artificial propagation for its product.²⁷ The safety restrictions that the dealers agreed to after the two nationwide typhoid outbreaks only made establishing a thriving industry harder.

Faced with bad publicity even in the years before the outbreak, oyster dealers began to self-regulate. "The New York City dealers are conducting business under permits issued by the Board of Health of the City of New York, granted only after the most rigid inspections," explained an Ithaca, New York, paper in 1909, "and are lending every assistance in their power to the health authorities to bring about the desired object of assuring people of the absolute purity of the oyster supply of New York City." The paper told readers, who lived well within the range for getting their oysters directly from the city, "There is no food product so thoroughly inspected both as to the condition and the environment or utensils, in which it is prepared for shipment to the retailer or consumer, as the oyster."²⁸ The 1911–12 typhoid outbreak proved that those efforts didn't work. It took another outbreak for regulations to be strengthened enough to prevent more of similar size.

The report on the 1911–12 outbreak identifies the two wholesalers involved only as New York Dealers "A" and "B." It is extremely likely that at least one dealers was located in the Fulton Fish Market because most of the dealers in the city were there at that time. A new outbreak in 1924–25 that originated on Long Island near West Sayville definitely included Fulton Market wholesalers for the same reason. This outbreak was discovered in Chicago, demonstrating how long oyster supply chains were. An inflated number of typhoid cases there was immediately traced to the consumption of raw oysters, which were traced to Long Island. Shortly afterward, similar outbreaks were recognized in New York City, Washington, D.C., and several other cities.[29] Chicago was the worst hit, with hundreds of cases and at least eleven deaths.[30]

The outbreak could be traced directly to a single spot on Long Island because most of the people who fell sick knew what kind of oysters they had eaten. Those oysters, branded Bluepoints, were gathered and prepared at a facility just outside of West Sayville. In Chicago, investigators determined that out of 1,648,860 oysters sold in the area at that time, 755,130 were Bluepoints. It also helped that many consumers knew they had eaten Bluepoints because they had asked for them by name, having a particular affinity for that brand.[31] In 1909, the firm's founder, Jacob Ockers, had 500,000 bushels of oysters planted at West Sayville and was the largest planter, dealer, and shipper on the East Coast.[32]

Because this outbreak was bigger than the one in 1911–12, government investigators had a harder time making definitive conclusions about its source. One possibility was melt ice used to keep the oysters fresh.[33] This would have explained why not every Bluepoint oyster shipped had been infected since some, but not all, of the ice used to keep them fresh may have been tainted. The Bluepoints Oyster Company bought oysters from all over the Great South Bay, sorted them, and branded the best ones. Besides tracing the outbreak from Long Island across the country, the government investigators noted a spike in typhoid fever cases in Long Island before the outbreak, but they couldn't pinpoint exactly which beds had been infected.[34] In this way, the increasingly complicated supply chain for oysters actually threatened the whole industry because suppliers couldn't just close off a single source and declare the problem solved.

Rightly alarmed that this Long Island–based outbreak would destroy the trust of the American public in oysters of all kinds, government officials and representatives from across the shellfish industry met in Washington in 1925 to hammer out new regulations to keep their products safe. The resulting suggestions included inspection of oyster beds, the plants where oysters were processed, and storage and shipping facilities.[35] The long-term health of New York's oyster industry only worsened over time. In 1926, both New York State and New Jersey banned the floating of oysters because puffing them up with water was deceptive to consumers.[36] The city's Department of Health closed the last oyster bed around the city for safety reasons in 1927.[37]

In the last installment of the Old Mr. Flood trilogy, Joseph Mitchell's character tells a long story about a horse who died after being fed too many oysters. In the middle of the story, Mitchell-as-Flood sidetracks from the main narrative to lament the passing of the oyster barges that were once lined up on the East River near the market:

> At the time I'm speaking of—in 1912—there were fourteen barges at Pike Street, all in a row and all paint as loud and bedizy and fancy-colored as possible, the same as gypsy wagons; that was the custom. George Still is dead, God rest him, but the business is there yet. His family runs it. It's one of the biggest and brightest shellfish concerns in the city; and it's right there in the old barge, head office and all—George M. Still, Incorporated, Planters of Diamond Point Oysters. Still's barge is the only one left, and it's a pretty one. It's painted green and yellow and it's got scroll-saw work all over the front of it.[38]

George M. Still's barge closed in 1949 when the city took over the pier where it docked.[39] Between 1920 and 1966, oyster production in the United States dropped by 50 percent.[40] Oysters, once the regular food of New York's masses, became seasonal food eaten by Americans primarily on holidays like Lent, leading up to Easter, and Thanksgiving.[41] Pollution did not completely destroy the industry in New York City, but it

FIGURE 10.1 Oyster barges near Fulton Fish Market. "Manhattan: South Street—Pike Slip," by Berenice Abbott, 1937.

Photographic views of New York City, 1870s–1970s, from the collections of the New York Public Library

transformed the provisioning chain in ways that nineteenth-century oyster dealers never would have imagined.

As was the case with various fisheries, Fulton Fish Market oyster dealers were not directly interested in any particular oyster bed. This meant they could always buy their shellfish from nonpolluted sources far from the city. In 1919, Fulton Market offered oysters from thirty-three different bays and harbors all along the Atlantic coast, ranging from Prince Edward Island in Canada down to Florida.[42] "The oyster wholesalers in

New York were the unseen powers in the Staten Island oyster business," explained George Hunter to Joseph Mitchell in 1956, talking about the state of oystering there after the beginning of the typhoid fever scare. "They advanced the money to build boats and buy Southern seed stock. When the typhoid talk started, most of them decided they didn't want to risk their money any more, and the business went into decline." Sandy Ground, the African American settlement on Staten Island where George Hunter lived, never recovered.[43]

Even if dealers could pivot to sourcing oysters from other beds, by the 1920s, a huge number of the beds on the East Coast had become depleted.[44] Decentralization hurt the oyster wholesalers in Fulton Market too. One reason for the consolidation of New York City dealers in Fulton Market is likely the consolidation of other wholesalers outside city limits. For example, Jacob Ockers's Bluepoints Oyster Company had a factory on Long Island that processed and shipped 23,000 barrels of oysters in 1899 to foreign countries and every section of the United States and its own oyster grounds across Long Island and New England, which made the firm completely vertically integrated.[45] By the early 1930s, most of the shucking and processing was done near the oyster beds on Long Island. The wholesalers at Fulton Market were simply a link on the long supply chain.[46]

During the early 1920s, the first Pacific oysters were seeded from Japanese stock between Northern California and British Columbia. These would eventually prove to be stiff competition for all the East Coast varieties.[47] By the middle of that decade, other oyster distribution hubs emerged in Omaha, Denver, Minneapolis, St. Paul, Milwaukee, Chicago, St. Louis, Memphis, and Nashville, at a time when the supply of oysters was decreasing and the price was going up. The oyster industry had been planting seed oysters on natural beds for years, but now those beds were becoming completely unusable for a variety of reasons, especially depletion.[48] The fish hatchery at Cold Spring Harbor was one of a number of places around the state that figured out how to raise oysters in the lab, then plant them in the water when they were big enough to survive, but there were far fewer places to raise oysters safely. Pollution in the Long Island Sound severely limited where oysters could be raised on both the Connecticut and New York shores during the mid-twentieth century.[49]

In September 1938, a particularly bad hurricane struck eastern Long Island on its way to doing even more damage in New England. The storm devastated the oyster industry all over the Long Island Sound by moving tons of sand onto the oyster beds. Many oysters were picked up from one bed by the storm and transferred to beds with different owners, devastating small oystermen on the wrong end of this displacement.[50] In the long term, the damage created a new inlet near the protected waters south of the Hamptons, letting in more predators and increasing salinity. None of this was good for raising young oysters.[51] It became impossible to set oysters in the Great South Bay from that point forward.[52]

To make matters worse for the long-term health of this industry, the Long Island Sound was not particularly good for oyster culture to begin with. The waters were too cold to grow young oysters through aquaculture. Most had to be planted elsewhere and moved later.[53] By the 1940s, growers had to transfer their oysters at least three times over their five- to nine-year growth period in order to have a marketable product at the end. This had been done during the nineteenth century, but not so many times. This was a sign that even the beds on Long Island were beginning to be depleted.[54] As was the case with the beef industry, substantial human intervention was necessary for production to meet demand. Then during the 1950s, two different oyster parasites devastated the local industry. The results of all these difficulties were visible in Greenport, Long Island, for example, where decades later the remains of two giant oyster canneries that went out of business lingered near town.[55] In the 1970s, oyster production in the United States was only 1 percent of what it had been in 1910.[56]

At the height of the oyster boom of the late nineteenth century, oysters were both hyperlocal and harvested in staggering numbers at the same time that wealth inequality was growing in the city. In 1880, 765 million oysters were harvested from the waters around New York City. That number dropped to 300 million in 1907.[57] In the late 1930s, 10 million pounds of shell oysters and a million pounds of shucked oysters passed through Fulton Market each year.[58] Most of them no longer came from the surrounding waters. During the 1940s, one of the Fulton Market oyster traders, Thompson & Potter, tried returning to Raritan Bay and other places nearby. They got about 200 bushels in two years. Most of the oysters were large, ten- to fifteen-year-old specimens that proved unsellable.⁵⁹ Thanks to pollution, what remained of the Long Island oyster industry became New York City's local oyster industry. The trade there eventually recovered somewhat, but it would never be able to feed the city's masses again.

At the beginning of "The Bottom of the Harbor," Joseph Mitchell wrote, "The bulk of the water in New York Harbor is oily, dirty, and germy. Men on the mud suckers, the big harbor dredges, like to say that you could bottle it and sell it for poison." Nevertheless, there were still baymen who earned a living fishing along the outer edges of New York Harbor in Mitchell's day. Toward the end of that same essay, one of them noted that the shad caught around Staten Island and in the Hudson tasted like kerosene. Considering the state of the water, this is not surprising. What is surprising is that anyone would eat them at all.⁶⁰ Fishermen had been among the earliest advocates for government efforts to fight pollution around New York City, but they didn't succeed. Both New York and New Jersey either banned or advised against the consumption of striped bass, eel, and a few other fish caught around New York Harbor during the mid-1970s.⁶¹ Around the same time, there was a mercury scare based on the federal government's tests of swordfish that led to a 50 percent drop of sales at the Fulton Fish Market across the board.⁶² Even today, the water in the Hudson River is safe to swim in thanks to sewage treatment plants, but the fish that swim there remain unsafe to eat thanks to cancer-causing PCBs.⁶³

Other forms of industrial pollution also continued to be a problem. "It's hard to detect [fish from oily waters]," the buyer for the Grand Central Oyster Bar explained in 1976. "You have to stick your fingers into the gills and smell the oil." During the 1980s, signs in area fish stores bragged about how their fish came from faraway places with cleaner water than New York.⁶⁴ Some Fulton Market dealers bought tainted fish from the Hudson and sold them for much more, claiming they came from the Chesapeake Bay.⁶⁵ For all these reasons, the reputation for fish as a healthy food suffered permanent damage. Consumers who still ate fish came to accept that consuming seafood was always at least a little dangerous. People started caring about the stories behind their seafood as a way to assuage the fear surrounding their decision to keep eating it.

This was especially true of oysters. The difference between mercury poisoning swordfish and typhoid poisoning oysters is that the consumer will know rather quickly if they get typhoid, and that ailment can be directly traced to a single meal. A 1971 study produced by the Woods Hole Oceanographic Institute explained how successful twentieth-century oyster producers dealt with this risk:

> The most impressive advances in oyster culture have occurred and will probably continue to occur in areas such as Long Island Sound and on the Pacific Coast where vertical integration of the industry is maximized. The same company produces or purchases its own oyster seed, plants them on its own bed, maintains the beds, harvests the oysters, and in most cases processes the oysters for market. Under this system, the producer has exercised complete control throughout, and it is imperative that he assures the quality of his product in order to make a profit on his investment.[66]

How do you know you aren't eating an oyster that was taken from polluted water? Every oyster sold for food in the United States has a tag associated with it. The Food and Drug Administration requires restaurants that serve oysters to keep those tags on file for ninety days in case they need to trace the supply.[67]

The shorter the provisioning chain, the less there was to go wrong along the way. The shorter the provisioning chain, the less of a role existed for the wholesale dealers at the Fulton Fish Market. The history of the one restaurant in New York City that remained famous for its oysters during the twentieth century is emblematic of both the decentralization of the industry and its increasing emphasis on quality. The Grand Central Oyster Bar opened in 1913 as part of the Grand Central Terminal Restaurant, the main restaurant at the then new transportation hub Grand Central Station. Its creamy oyster stew had become an iconic dish by the late 1930s. Besides tasting good, the stew came with a little show, as it was cooked right there at the table. In the twentieth-century version of waiting for the ferry, commuters could also grab a snack while standing as they waited for their train home.[68] However, the main menu at the restaurant was standard order continental (in other words, French) cuisine. It remained that way until 1974, when the restaurant's new owner revamped the menu to emphasize its seafood.[69]

Therefore, the Grand Central Oyster Bar only highlighted a vast variety of raw oysters very late in the twentieth century. By 1986, the average number of types of oysters available there was between six and twelve.[70] "When I first came to the Oyster Bar as the fish buyer," recalls the executive chef, Sandy Ingber, "my only source for oysters was the Fulton Fish Market. And there would be times that I would be hard-pressed to find six varieties to put on the menu." When the dealers there had a monopoly over the city's oyster market, they had no reason to care. However, Ingber writes, "Since the late 1990s, there's been an explosion in producers. Now, you might find as many as thirty-five oysters . . . on offer on any given day."[71] Pollution was a problems everywhere, but beds that were constantly monitored and overnight air delivery made it possible for restaurants throughout the greater New York area to serve live oysters from all over the world 365 days per year.[72] The closest thing to local oysters came from Long Island.

In 1883, most oysters eaten in New York City came from local waters. Bluepoints from Long Island came from farthest away.[73] In 1925, wholesalers in the Fulton Fish Market could still buy their oysters from far out on Long Island or bring them in from out of state, but when more people decided that they didn't want to eat oysters anymore, there was nothing those wholesalers could do. Unlike in the Chesapeake Bay, the oysters on Long Island during the twentieth century were cultivated on private plots, leased from the state for purposes of aquaculture.[74] Private plots had been usual in nineteenth-century New York City, but demand and knowledge were not enough to make these operations profitable a century later, until the post-1990 New York oyster renaissance.[75] As the restaurateur Danny Meyer explained in 1986, when the Union Square Cafe was brand new, "I've learned a restaurant secret about serving raw oysters: When we have just one kind on the menu there are very few orders, but if there is a selection of several different kinds, and people can compare them, we get a rash of orders."[76] Depending entirely on oysters from local waters was no longer a good way to attract customers.

Despite the pollution, many people in the greater New York area still ate shellfish harvested from around New York City. The key to making this work was that they knew exactly when oysters were most likely to be clean, despite their location. As Mitchell explained, "They watch the temperature of the water to make sure the oysters are 'sleeping,' or hibernating, before they eat any. Oysters shut their shells and quit feeding and begin to hibernate when the water in which they lie goes down to forty-one degrees; in three or four days they free themselves of whatever germs they may have taken in, and they are clean and safe."[77] Knowing that required direct knowledge of shellfish that everyone else had lost by the mid-twentieth century because the successful provisioning system built up through Fulton Market had completely separated the vast majority of New Yorkers from the sea life all around them. By that time, longer supply lines, some through the Fulton Fish Market and some not, had begun to take the place of local seafood.

11

BUYERS

During the nineteenth century, when consumers bought only a limited number of fish species, the names of those fish didn't matter much. As long as what they bought tasted like what they thought of as fish, consumers were happy.[1] In the twentieth century, as the number of fish species sold at the Fulton Fish Market grew, the names mattered more. Many similar-looking fish have different names. Angler fish are also called monkfish, goosefish, bigmouth, and lotte.[2] Fish that look completely different sometimes have the same name. There are at least seven completely different kinds of fish around the world that are generally known as moonfish. Some look at least a bit like moons; others don't.[3] The dealers who worked at the Fulton Fish Market had to standardize the names of fish in order to be able to price and sell them. The buyers, on the other hand, could call those fish anything they wanted as long as their customers didn't know what they were or couldn't taste the difference from what they thought they were eating.

In 1897, the *New York Times* wrote about a fish retailer in Little Italy named Alphonso Masuci who had acquired a particularly large "tunny fish." Today we would recognize it as a bluefin tuna, common in the Mediterranean at that time. "It is extensively used in the Italian countries," wrote *Scientific American* in 1877. "It is pickled in various ways, boiled down in soups, and made into pies, which are thought to be very excellent and possess the valuable property of remaining good for nearly two months."[4] When Italian immigrants arrived in New York City in large numbers around the turn of the twentieth century, they brought their taste for "tunny fish" with them. "Strange to say," wrote George Brown Goode in 1884, "although highly prized in the Old World from the time of the ancient Romans to the present day, they are seldom, if ever, used for food in this country."[5] Once left to rot on the shores, tuna began to appear in Fulton Market when dealers knew there would be buyers.

Masuci bought a 900-pound bluefin tuna at Fulton Market for $9. "To the Italian the tunny fish is the fish of all fish," the *Times* reported. "They have a way of preparing it with olive oil and a peculiar sauce the memory of which will make a man's eyes moisten and his mouth water." When the fish arrived at his store, "a changing group made up of old women, old men, young girls and small boys with bulging eyes worshipped at the shrine. The old people spoke of fish in their day in far-off Italy, and revealed the secret of the wonderful sauce. To this last, the young girls paid careful attention."[6] Because bluefin tuna have incredibly long migration routes, this was likely a particularly notable specimen of the exact same species Italians like Masuci ate back in their homeland. Undoubtedly, it was close enough for the customers, who bought parts of it to prepare them he same way.

For most of the nineteenth century, the dealers in Fulton Market called bluefin tuna horse mackerel.[7] While Pacific albacore tuna found a market as canned tuna fish, up until the 1970s bluefin tuna was often discarded by northeastern fishermen even though they recognized that it was good to eat.[8] Because the fish are both extraordinarily large and very energetic, they often damaged the nets that caught them, thereby letting smaller fish get away.[9] Sport fishermen on the Pacific Coast made large tuna popular among rich people with plenty of time and money to hunt them down.[10] The idea of commercial fishermen deliberately seeking them out would have been foolishness. Even Italians who loved tuna in the old country were not enough to maintain a market in New York because cutting up so large a fish and getting it out fresh to all but the most local buyers would have been very difficult.

FIGURE 11.1 Interior view of the Fulton Fish Market showing scales, crates, and baskets. Walter Albertin, *World Telegram & Sun* photo, 1954.

Library of Congress, Washington, D.C.

The existence of the Fulton Fish Market depended upon buyers for the fish sold there. When those buyers moved from retail to wholesale over the course of the nineteenth century, their relationship with sellers became even more important because the pool of buyers became much smaller. Both sides had to stay happy for the provisioning chain to remain effective. The first buyers were mostly other dealers, but this evolved from dealers around the country to mostly those who sold fish in New York's other markets. By the turn of the twentieth century, many had their own stores. Restaurants, too, became important purchasers of fish, especially the few that actually specialized in seafood. After 1920, as grocery stores became increasingly common, a new link in the provisioning chain developed for companies that processed fish for freezing and distribution through those channels so that consumers would be able to cook fish more easily. In response, fish retailers and the market that served them began to specialize in new species.

This meant convincing consumers to eat fish for which markets had not previously existed. Trash fish are species taken up in nets that the fishermen didn't want because they had too few buyers. However, the number of buyers for some species, like skate, improved over time. "New England fleets ship trash fish in small quantities to Fulton Market," wrote Joseph Mitchell in 1947, "where they are sold to two dissimilar groups: buyers for luxury hotels and restaurants, and proprietors of little one- and two-bin fish stores in Italian, Spanish and Chinese neighborhoods."[11] During the 1970s, the Grand Central Oyster Bar told all its suppliers that it would start buying wolf fish. This one buyer was enough to ensure the species ceased to be trash fish.[12] Catering to these kinds of customers by providing expensive fish in smaller amounts soon became the future of the Fulton Fish Market.

While Fulton Market had depended upon hotels for much of its business, the Fulton Fish Market depended upon restaurants. New restaurants specializing in seafood and shellfish were part of the mystique that accompanied the development of New York City's Broadway theater district between 1890 and 1920. Rector's, an import from Chicago, opened in 1899. It was the first in a series of restaurants that came to be known as "lobster palaces." Rector's began life in the middle of Manhattan and was totally electrified, which gave it the same aura as the nearby theaters. It also had the first revolving door in New York City. The menu was not exclusively fish, but its fish was the kind of menu item that made a place famous. The financier "Diamond" Jim Brady, for example, became obsessed with fillet of sole while traveling in France, and Rector's became the first restaurant to serve it in the United States in the French manner in order to please him.[13]

The Manhattan Oyster and Chop House—known as Jack's after its owner, Jack Dunstan—started life as an oyster bar. While a chop house obviously served steaks, it also had mussels, clams, and its specialties: lobsters and lobster fat on toast. These large restaurants catered specifically to the upper classes who were interested in showing off their newfound wealth, but still got customers from all social classes.[14] The George Still Company, which supplied oysters from a barge anchored near the Fulton Fish Market, actually owned its own seafood restaurant on Third Avenue while the oyster business was still booming. The wholesale business outlived the restaurant, which may explain why that kind of vertical integration was not particularly common.[15] Lobster Newburg—a rich dish that includes lobster, butter, eggs, and cream—originated at New York's most famous restaurant, Delmonico's, around this same time.[16]

The restaurants immediately surrounding the Fulton Fish Market always had the advantage when it came to fresh fish. As the neighborhood emptied of residents, there still had to be places where people who worked in the market could eat during their long days. "You would think [the men in the market would] get sick of fish, but they don't," explained a patron at Sloppy Louie's to a writer for *Newsday* in 1966.[17] Inevitably, those restaurants specialized in seafood. People who worked with fish and oysters all night long actually liked eating them too. Initially, the most famous of these places was Sweet's, probably because of its longevity. A. M. Sweet opened his restaurant in 1845 on Fulton Street, in Schermerhorn Row, directly opposite the market. It was known primarily for its lunch service, which attracted businessmen from all over Lower Manhattan.[18] In the nineteenth century, like many other restaurants, it sold dishes of all kinds. Only during the early twentieth century did the restaurant start to specialize in fish.[19]

Sweet's cooked its fish simply because the freshness of the fish made it special. Customers thought the portions were huge, and the service was always very slow. Oysters were a mainstay until the quality available at the market declined during the late twentieth century.[20] A 1938 review offers a glimpse of a restaurant that was already feeling its age:

> Electric light was substituted for gas for some compelling reason, but the same old floors are still bare and are swept by colored waiters with the same old broom-corn brooms. No fancy carpets or vacuum-cleaners for Sweet's!

> The building was once a hotel, and, hard though it be to realize it now, they say it was a swell house back in the days of U.S. Grant. The restaurant is up one sweeping flight of stairs, for it was the dining room of the hotel. The old building sags under the weight of years, and the rooms of the once-swank hotel are vacant and bare, with unused gas jets sticking out from the walls.[21]

Sweet's eventually became a popular destination for tourists, who would visit the market during its morning operations and go to the restaurant for their afternoon meal.

Sloppy Louie's, the restaurant that Joseph Mitchell made legendary, was also located in Schermerhorn Row. Louis Morino, an Italian immigrant, bought the restaurant that had been known as Sloppy John's in 1930. One of the regulars from the market started calling it Sloppy Louie's to irritate the immaculate Morino, and when that moniker became popular he decided to lean into it. "Sloppy Louie's is small and busy," wrote Mitchell in 1952. "It can seat eighty, and it crowds up and thins out six or seven times a day. It opens at five in the morning and closes at eight-thirty in the evening. It has a double door in front with a show window on each side. . . . There are mirrors all around the walls, four lamps and three electric fans with wooden blades that resemble propellers hang from the stamped-tin ceiling." Customers could choose from twelve long wooden tables, but because they were communal and the restaurant was usually crowded, people from all walks of life might find themselves dining next to one another. The windows of the top four floors of the building that contained the restaurant had been boarded over for many years when Mitchell wrote about the place in 1952. The abandonment of those floors was the nominal subject of the essay "Up in the Old Hotel."[22]

Of course, Sloppy Louie's specialized in fish. "The fishmongers use Louie's as a testing kitchen," Mitchell wrote. "When anything unusual is shipped to the market, it is taken to Louie's and tried out. In the course of a year, Louie's undoubtedly serves a wider variety of seafood than any other restaurant in the country."[23] This is how the dealers gained knowledge about what new species they might sell were good to eat. Morino's bouillabaise, which contained a vast array of sea creatures from the market across the street and which he insisted that no housewife could ever

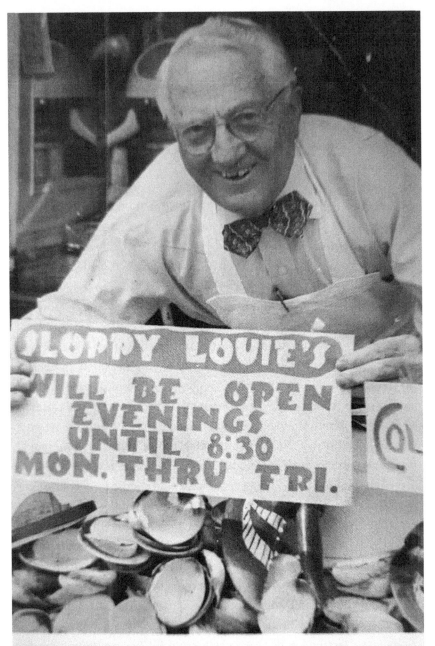

afford to duplicate, was the house specialty.²⁴ In its last years, the restaurant's promotional literature bragged, "There is a quaint rusticness found here amidst the commotion off the nation's most active financial district.... The service is fast, simple and direct. You can eat-and-run in keeping with the demands of a busy schedule, or are most welcome to linger, even revel in an environment reminiscent of days gone by."²⁵ The place was filled constantly not just with men who worked at the Fulton Fish Market but also with people like Joseph Mitchell, who came because the restaurant was so closely associated with the market itself.

Restaurants were particularly concerned about the size of fish they bought because they had to control their portion sizes. Fillet houses felt the same way because they could get more fillets from larger fish.²⁶ Filleting, like the beef-processing industry in Chicago, allowed processors to sell the parts that consumers didn't want at a substantial profit. Much of the surplus fish went into processed canned cat food, glue, and fish meal. The whole sequence saved on transportation costs. Once fillets were packed in tins or boxes, handling them became more sanitary. Because the packaging could be branded, consumers came to trust fillets more than just some undifferentiated fish with obscure origins.²⁷ Filleting a fish turned the corpse of an irregular creature into a regular thing that could be sold in discrete, uniform portions to institutional buyers, especially frozen fish companies.

The Bay State Fishing Company of Boston pioneered the processing of fish fillets in 1921.²⁸ The idea was to take the difficult and tedious work of preparing a fish out of the hands of the housewife in order to increase fish consumption in the home. The fish were cut, brined, wrapped in parchment paper, and shipped to retailers for rapid sale. The practice

FIGURE 11.2 Louis Merino of Sloppy Louie's in a promotional postcard from sometime in the 1950s.

Author's collection

spread quickly.[29] A little later, the first fillet plant appeared on Long Island. Its owner bought fish straight off the boat, processed them, and shipped them off to supermarkets, bypassing the wholesalers.[30] Inevitably, fillet men entered Fulton Market itself by the mid-1930s.[31] By the 1950s, there were fillet companies and fillet men throughout the surrounding neighborhood. For example, Smitty's Fillet House at 100 South Street had a nine-man team of shad filleters that prepared 17,000 pounds of shad in just one week during the 1961 season.[32] They sold to wholesalers, who in turn sold the processed fish to hotels, restaurants, and markets.[33] Mass-produced fillets made by machine in places like New Bedford, Massachusetts, used chemicals such as sodium tripolyphosphate to keep the fish fresh and maintain its appearance. Fulton Fish Market fillets were not just preservative free, they included different fish (like shad) that simply couldn't be filleted by machine.[34]

The unofficial uniform of the fillet man was a long rubber apron and white cotton gloves. He carried a seven- or eight-inch-long carbon steel knife that was slightly curved and just a little flexible to deal with fish that had complicated bone structures.[35] In 1986, the fillet houses were across the street from the market. They opened at 1:30 a.m. Fillet machines began to appear on factory trawlers when those massive ships were first built during the 1950s, as processing the huge hauls they caught was one part of the efficiency of those vessels.[36] The fillet houses around the Fulton Fish Market still employed hand labor because humans could do the work better. Unlike the fillet houses that depended upon bulk, they were counting on quality.[37] Buyers who came to get the freshest seafood possible for their retail fish stores could stop at a fillet house in the neighborhood to pick up whatever preprocessed fish they needed on their way to South Street to haggle with the salesmen there.[38]

Fillets could also be sold directly to consumers for baking and grilling. They were an important transition point between fish as a natural resource and fish as a commodity because they were not immediately recognizable. Fillets were initially advertised as being "sweet and odorless," two qualities seldom associated with fish. Unlike fresh fish, fish fillets could last indefinitely because they were frozen more successfully than previous fish products thanks to Clarence Birdseye's innovations in this particular area of the industry. Perhaps most importantly, cutting

FIGURE 11.3 Vinny, a Fillet Man, 1982. He normally wears the standard long rubber apron, but in this picture has no white cotton gloves.

Photo by Barbara Mensch, in *A Falling Off Place: A Unique Story of Lower Manhattan*, Empire State Editions (New York: Fordham University Press, forthcoming)

fish into fillets made it look more like its main competitor on many American dinner tables, meat. The fact that fillets were so easy to prepare only made them more appealing.[39] By the second half of the twentieth century, firms like General Foods had developed long cold chains to assure that their preserved fish would stay fresh all the way from the point of catch to the consumer's kitchen.[40]

Starting in 1950, fillets became an intermediate point in the creation of fish sticks. The primary way that homemakers cooked fillets was to bread them in batter and fry them. Fish sticks were frozen fish "steaks" that had been cut, breaded, fried, and then refrozen.[41] In the same way that fillets meant women didn't have to process a whole fish, turning fillets into fish sticks saved even more work. Sales took off instantly. As Jane Nickerson of the *New York Times* explained in 1954, "Coming in from Canada, Iceland and Norway, these fillets depressed prices of domestic fillets to such an extent that a year ago certain members of the trade were discussing among themselves the advisability of seeking a high tariff on foreign imports. But now the industry is eager for as many fillets as possible for processing into sticks." The species didn't really matter when it came to fish sticks since they all tasted like batter. The freshness didn't matter either, since they were all frozen.[42] The greatest advantage of the fish stick was the ease with which homemakers could get them onto the table.[43] Because Canada, Iceland, and Norway were also the countries that ran the giant super trawlers off of America's coasts, the efficiency and volume of the operations translated into cheap, convenient food. Another area of growth for the seafood industry during the second half of the twentieth century was canned tuna. This provisioning chain consisted of large canners buying directly from fishermen, leaving intermediaries out from the very beginning.[44] Competition from large-scale fish processors of all kinds led to a steady decline in volume at the Fulton Fish Market beginning in the mid-twentieth century and continuing for the rest of its time in Lower Manhattan.

With little freezer or cold storage capacity, the Fulton Fish Market couldn't compete on the basis of price. Instead, during its final decades, it became an important place for fresh fish of many kinds. Increasing variety may not have been a deliberate decision. The wholesalers at the market sold a wide variety of fish because in New York there was demand, which started with members of ethnic groups who traditionally ate more exotic seafood than other Americans did. "Fifty years ago, New Yorkers ate little more than codfish, flounder and shrimp," explained the writer Philip Lopate when he visited the market in 2001, "but their tastes have grown more adventurous and cosmopolitan as the city's demographics have changed. Philippine, Thai, Korean, West Indian, Dominican, Haitian and Indian cuisines have all widened the fish-eater's palate, just as Japanese sushi has taught people to appreciate the once-disdained tuna."[45] Besides changes in the customer base, another reason for new fish appearing in the market was that the U.S. government's conservation measures designed to support domestic fisheries forced the wholesalers to seek out whatever fish was available, and much of that came from farther away.[46]

By 1987, the market handled approximately 450 different varieties of fish.[47] In 1880, the market had handled 44 species over the course of almost an entire year.[48] This increase in variety was only possible because of two kinds of improvements that had already changed Fulton Market one hundred years earlier: transportation and refrigeration. By the mid-1970s, obtaining fish by airplane was only slightly more expensive than getting it via truck.[49] Even if these fish weren't kept in cold storage at the market, they invariably were preserved through mechanical refrigeration at some point on their long journey from all over the world. Together, refrigeration and air transportation allowed the Fulton Fish Market to defy seasons and the passage of time by delaying the inevitable decay. Wholesale dealers could then serve immigrant New Yorkers with seafood direct from their home countries.[50] Other new fish introduced in this era included golden king clip from Chile, Alaskan pollock, and two specialties from Hawaii: moonfish and mahi mahi.[51]

Chefs and restaurants also played a role in this development, popularizing what had once been ethnic seafood for other locals. Squid is a good example of this phenomenon. "Back in the late '70s you couldn't give squid away," explained a man whose job an Associated Press reporter described as "squid pusher" in a 1987 story. "Now, a lot of restaurants have it on their menu. But most are calling it calamari (the Italian word for squid) because when you say squid people have a vision of Captain Nemo being dragged down into the deep by a monster shooting ink." In that same story, this information officer for the Fulton Fish Market specifically credited restaurants serving Chinese and Italian food for the increase in squid's popularity.[52] There is a certain poetic justice in that unlike other kinds of seafood, the squid's numbers have increased since the advent of industrial fishing, in large part because fishermen have decimated the population of the fish that are its natural predators.[53]

Another fish that began to appear regularly in the Fulton Fish Market around this time and reflected the new diversity of species was the bluefin tuna. Developing this particular market required that fish be able to travel by airplane. As early as 1940, the Fulton Fish Market already received a substantial amount of product that way, much of it from the Pacific Northwest.[54] In 1970, Boeing introduced the 747 jet airliner. Japan Air Lines experimented with special containers to transport Atlantic bluefin to Japan safely shortly thereafter. Norwegian innovation that created leakproof containers and a combination of both artificial refrigerant and ice made long-haul flights for fish possible during the same decade. The global market for other fish quickly followed. Much of the fish that made Gloucester tuna fishermen rich during the 1970s went to Japan, but before long the American market became equally lucrative.[55] Bluefin tuna began to appear at Fulton Fish Market regularly once the sushi boom began during the 1980s. However, fish traveling by air completed the Fulton Fish Market's transformation into nothing more than an entrepôt for product entering the New York area.

Unlike oysters, fish don't have tags that determine their provenance. When processing fresh fish into fillets became common early in the twentieth century, it formed another link in the provisioning chain. The more links in that chain, the harder it became to tell where a fish came from and in this case, what that fish really was. Fraud is a surprisingly difficult term to define. Dealers could call a fish something that it wasn't in order to charge more money for it than they would otherwise get. They could also call a fish something that it wasn't in order to make it sell faster. Because so many species of fish went under many different names, such deceptions were next to impossible to regulate. The fact that the wholesalers at the Fulton Fish Market treated different customers differently made these practices even harder to detect.

Fresh fish is a product of highly variable quality. Even a good specimen will go bad quickly without proper preservation. Those customers familiar with the culture and the operation of the Fulton Fish Market were likely to get high-quality fish. Those not familiar with that culture could end up with fish that might otherwise be headed for the garbage pile because it was no longer fresh. The standards of freshness for fish became higher as the technology of refrigeration and transportation improved. As late as the 1970s, the freshness of the fish available at the Fulton Fish Market varied by the day because there were fewer networks supplying seafood to Americans who ate substantially less of it than they do now. The industry didn't grade fish, so buyers were to a great extent at the mercy of wholesalers, and the wholesalers simply couldn't offer top-of-the-line, fresh-off-the-boat seafood to everyone. Lucky for restaurateurs and retailers, most of the customers couldn't tell the difference.[56]

Both the appearance of fish sticks and the difficulty of identifying particular fillets were sure signs that fish was becoming a commodity during the late twentieth century. So was the practice of spraying or dipping fillets to hide decay by improving their color.[57] Highly processed, well-preserved fish could be sold, traded, and stored based on supply and demand that varied considerably over time. Wholesalers and buyers had a greater opportunity to create value for fish where none had been present before because they had more time to market their product. Filleting it was one way to create value. Preparing it well in an expensive restaurant was another. Few Americans could differentiate a pollock from a whiting, but the expertise held in and around the Fulton Fish Market created this second alternative by providing the freshest fish possible.

Long-standing buyers learned to understand the language of the wholesalers. Dealers at the Fulton Fish Market had their own terms for some of the fish sold there. A black-barred hogfish might be labeled parrotfish. Starting in the 1980s, red porgies were often labeled as red snapper. Groupers, a particularly large group of related fishes, were often misidentified—not through malice necessarily, but because they could be so hard to distinguish from one another. Snappers were a particular problem during the cajun/creole cooking craze when many New York restaurants and their customers wanted to try New Orleans chef Paul Prudhomme's blackened redfish, but the market actually provided many different fish under that name.[58] What mattered was whether people liked the fish they were eating, and in New York the price paid for a product like fish could be a status symbol. Since nobody else could tell the difference, did anybody really lose if the species they ate was slightly different than advertised?

The newfound emphasis on variety at the Fulton Fish Market was a rebellion by largely ethnic institutions against the homogenization of food in general during the late twentieth century. Children who grew up on fish sticks thought of fish as a single undifferentiated mass, but this was the era of not just frozen fish but frozen food of all kinds. For many Americans, the convenience symbolized by the microwave oven trumped the taste of whatever dishes they happened to be eating. The market survived on the basis of selling a large variety of specialty products to New Yorkers who, for whatever reason, were willing to pay more for fish that most Americans didn't appreciate as much. "About 20 top-notch chefs come down here," explained the chef Jean-Georges Vongerichten to a reporter from *Nation's Restaurant News* in 1988. "They're part of a new breed who are buying direct; very few of the older chefs do."[59] These chefs also shopped for fish through new channels, which took business away from the Fulton Fish Market as the other options grew.

When the now legendary seafood restaurant Le Bernardin opened in 1986, its staff began working directly with the purveyors at the Fulton Fish Market. Gilbert Le Coze, the first head chef there, introduced black sea bass, skate, monkfish, and sea urchin to the United States by creating a market for it among fishermen and the wholesalers who bought from them. His staff called fishermen and told them how to care for the fish—how to chill monkfish, for example—because the market that Le Coze developed paid them to do so. The problem for the Fulton Fish Market with this kind of business was that restaurants like Le Bernardin started to buy much of their product directly from the fishermen themselves, bypassing the market entirely.[60] Letting Gloucester handle most of the fish sticks was a small sacrifice since Lower Manhattan could never cope with the volume involved, but separate provisioning chains for high-quality fish that serviced restaurants in their traditional territory became a serious threat to the bottom lines of many wholesalers.

There is now a whole specialty seafood industry designed to improve the quality of the product by shortening the chain, cutting middlemen out entirely. This kind of decentralization hurt the wholesalers when they congregated in Lower Manhattan, but it would become an important survival strategy for them in the years after the market moved to the Bronx.

Because high-end restaurants like Le Bernadin instilled customers' appreciation for particular fish, quality proved key to the market's survival. Grocery stores and fillet houses that depended on mechanical cutters (who were mostly located far away from Lower Manhattan) served fish to the masses. The wholesalers at the Fulton Fish Market sold fresh fish to restaurants and retailers whose customers appreciated that fact. "I sell to Sardi's, Fraunces Tavern, a lot of your big name restaurants," a crab dealer named John Catena told a correspondent for the *Atlantic* in 1975. "Frozen soft crab is now a tremendous industry, bigger than fresh, growing all the time. It takes care of glut periods. But it can also disguise a poor product, and I would hate like hell to see the fresh crabs go."[61] Selling high-quality, fresh seafood fetched higher prices and remained profitable even as the overall volume of fish going through the market dropped year after year. This business was lucrative enough that it could even survive the cost and inefficiency imposed upon it by the criminals who controlled so many of its operations. This strategy was popular enough that it also became a prime talking point in small wholesale dealers' prolonged resistance to the forced relocation of the market to Hunts Point in the Bronx.

12

THE CULTURE OF THE FULTON FISH MARKET AND ORGANIZED CRIME

Barbara Mensch moved to the neighborhood around the Fulton Fish Market in 1979. She soon began to photograph the people who worked there. In order to earn their trust, Mensch showed up night after night, photographing the market as it was rather than trying to romanticize it.[1] She went on to publish two books of her photographs, accompanied by interviews and descriptions of the people she met.[2] Mensch's portraits reflect her interest in the individuals who worked there, and her interviews are practically unique in the history of the market because few visitors had cared to talk to the workmen who made the market go over its many years of operation. Her work reveals the culture of the Fulton Fish Market in ways that simply are not available for earlier periods in this institution's history. One reason that earlier visitors found the culture of the place impenetrable was that they often talked to the wrong people.

"There's an old notion in Fulton Market," explained one of the other characters in the last of Joseph Mitchell's three "Old Mr. Flood" stories, "if you want to know a fishmonger's financial standing, don't bother Dun & Bradstreet, just look at him–if he looks like he's been rolling around in some gutter, his credit is good."[3] The richest of the wholesalers was known for his "particularly corroded coveralls" and was so cheap that he once hired an accountant to investigate whether his nephew was

FIGURE 12.1 "Fulton Fish Market in Session," c. 1936.

South Street Seaport Museum, New York, NY, gift of Vincent J. Tatick, Joseph H. Carter, Inc.

stealing from him.[4] This explains why the class politics of the Fulton Fish Market were usually tricky. Both dealers and their employees wore the unofficial uniform—duster, straw hat, big rubber boots—so it could be very difficult to tell them apart. Wholesalers were often described as "dealers" in the newspaper stories about the market, and often named. Other workers were usually anonymized because they were seldom deemed interesting enough for readers to care about them. Reporters (with the notable exception of Mitchell) mostly went to Fulton Market to write about fish, not people.

The gender politics of South Street were more obvious to those observers who looked for them. Although there were a few women fishmongers by the late 1980s, the market was nonetheless an extraordinarily male-dominated environment throughout its history. In a world that had already begun to recognize the achievements of women, women reporters arrived at the market and started to write about this nearly all-male society. Suzanne Hamlin of the *New York Daily News* wrote in 1985: "Of the 1500 people who work at the market, almost all are men, men used to lifting, hauling and making instant decisions. Used to standing in puddles of water, islands of mud, used to dealing with the torpid heat of summer, the biting cold of winter.... The camaraderie is obvious, the action is physical, and the noise is constant: 'Watch your back! Watch your back, *pul-ease*, lady!'"[5] Mensch was constantly aware that the gender dynamics were one of the reasons it took a long time for her subjects to trust her, since male workers were less likely to tell their stories to a woman. Many of them felt the need to speak with her outside the market because they didn't want to be seen talking about their lives with an obvious outsider.[6] For most reporters, the gender situation had always been taken as a given, but now it became part of the thick description.

In the nineteenth century, the Fulton Market was controlled by the same kinds of men who dominated the rest of society—mostly native-born Americans of English and Scottish descent.[7] There were fishermen who moved down from Connecticut and accountants who invested in fishing boats, but the basic profile was white, Protestant, and male. Philip Lopate, writing in 2001, picked up where the twentieth century began. "Traditionally the fish market has been the province of Italians and Jews," he explained, "though in recent years, Koreans have entered it, as well as a few Portuguese and Greeks."[8] A salesman with twenty-three years of experience elaborated on this dynamic the year before the market moved: "The ownership hasn't changed much; it's Italian and Jewish. But the work force has changed; you have more Hispanic, African-American, Asian. There are no ethnic boundaries down here; money is the joining factor. Whoever has the money to pay for the fish, you're my buddy."[9] In family-owned businesses, management and labor tended to be of the same race and ethnicity, and the businesses that survived passed down from generation to generation, which meant that the level of diversity in the market evolved very slowly.

Despite a tendency toward stability, around the turn of the twentieth century, wholesalers who were mostly of English descent were joined at the market by owners who were Irish, Italian, and Jewish.[10] In the 1930s, the fishermen who served Fulton Market were much more ethnically diverse than the people who processed and sold their catch. They included British, Canadian, Scandinavian, Portuguese, and Norwegian sailors.[11] Inside the market, employment often depended on a man's literal family, as fathers tapped sons or nephews for jobs.[12] The market was strictly segregated by race until the the era of the civil rights movement, when the first African American workers were hired.[13] The sociologist William B. Helmreich grouped the workers at the Fulton Fish Market with those of other white ethnic institutions in the city like the building trades, the police, and the fire department.[14] He might also have included the mafia, which held a presence inside the Fulton Fish Market for decades.

"Back in Dewey's time, they were investigating the market," explained Bob Cantalupo, whose family owned the garbage firm that served the Fulton Fish Market, to Joseph Mitchell in 1987. (Thomas E. Dewey was New York's District Attorney in the 1930s and early 1940s.) "A man came in to talk to us to check our books and he said let me see your contracts. I said, 'We didn't have any.' He said, 'How can you do business in a place like the fish market?' I said, 'We handle the market and we handle maybe 200 independent accounts and we don't have a single piece of paper. It's all done this way. We make a price.'"[15] This kind of trust was the privilege of being part of the culture. It was also a good way to avoid scrutiny from law enforcement if you were doing anything illegal, but that didn't put the wholesalers or the people they did business with completely above the law.

In 1950, Joseph "Socks" Lanza received parole. An investigation of the circumstances surrounding his parole began soon thereafter, on the premise that since he was related to two important New York Democratic politicians by marriage, he might have used his connections to get out of jail early. Lanza was sent back to jail, but no evidence of abuse of power ever surfaced.[16] In 1957, an informant told law enforcement, "Lanza has always been the boss in the fish market. He is still the boss; if you send him back for ten years he will still be the boss." That quote comes from a single witness in Lanza's parole report, which was never supposed to be released.[17] It was anyway. Lanza's new parole board concluded, "Joseph Lanza, through his family and combination, continues to control the

Fulton Fish Market. Specifically, this operation consists of shakedown of loaders, labor racketeering and the cornering of the shrimp, scungilli (squid), halibut produce market."[18] Despite the political implications of the controversy, newspapers around the country credibly reported the allegation that Lanza still ran the market because organized crime news sold papers.

Whether these charges were true or not, it is definitely true that law enforcement never cracked down on the corrupt system in the market between the mid-1930s and mid-1970s. If Lanza ran such a well-oiled machine that he could have controlled it from his prison cell, there should have been more investigations and more charges. At the very least, any system that Lanza managed from prison could have essentially run itself because most actors at the market consented to its operation. That would explain how this arrangement continued even after its architect faded out of the picture.[19]

After Lanza left the scene, the mob's power still came from its influence over the loading and unloading of fish. The Genovese crime family controlled all six firms approved to perform these functions, and it assigned each one to a particular supplier. As a result, there was no competition among members of the cartel. Usually all the unloaders did was take boxes of fish out of a truck and put them on the ground, but this unskilled labor was the choke point at the market because anyone who hired anyone else to do this would incur the wrath of the mob. Every Friday night at a Genovese social club, the unloading companies paid the representative of organized crime in the market in exchange for the power he invested in them over the wholesalers. There was also another line of payments made to the union directly that fell under the broad category of Christmas presents.[20]

The mob's continued power at the market depended upon maintaining the poor refrigeration infrastructure there because of the highly perishable nature of seafood. Even a slight delay could ruin a wholesaler if their product spoiled. If you refused to pay off the gangsters who controlled the union that represented the unloaders, your fish would spoil out in the streets. If the people handling the fresh fish demanded a bribe of any kind, you paid them. If you didn't, you'd eventually go bankrupt.

If you spoke up against the system, the gangsters might slash your tires or threaten your safety—or even your life. There were sporadic investigations of individual actors, but one mafioso could easily be replaced with others. The system itself was never threatened because everyone involved could still make money.[21] This worked well enough that the mafia faded into the background for thirty years. While it drained some of the wholesalers' profits and hindering the efficiency of their operations, everyone involved—including law enforcement—accepted its presence.

Organized crime revealed itself again at the Fulton Fish Market during the 1970s. At first the wholesalers paid a private security service to protect their business from what remained of organized crime, but then the union convinced them to hire a security service that was controlled by the mob itself. Random thefts stopped. However, the payments then went to Carmine Romano, who was also a part of the Genovese crime family. Romano had been secretary treasurer of the union local since 1974. Some of the fish dealers paid $1,300 a week for security to the "Fulton Patrol Services" to make theft drop dramatically. When the example was set, this arrangement spread to all the other dealers. The wholesalers ended up paying about a million dollars a year to secure their $750 million-dollar-a-year operation.[22] The wholesalers also got labor peace at the market, which kept the operation running and thereby making money for everyone. Indeed, strikes only hit the Fulton Fish Market when external forces threatened this system.

The U.S. attorney began a new investigation of the union's Fulton Fish Market activity in 1979. In 1981, he charged Romano and several other union officials with racketeering. Other union members routinely stole fish from drivers—a practice known as "tapping"—and sold it through a company that one of them owned. Ten wholesale fish firms pled guilty to making illegal payoffs to the union.[23] Romano and his twin brother Peter were convicted of racketeering and barred from serving in the union again.[24] Because their brother, Vincent Romano, took Carmine Romano's place as the contact between the union and the Genovese crime family, law enforcement had additional work to do.[25]

In 1982, when federal prosecutors charged Romano again, they alleged, "Organized crime controls every facet of the Fulton Fish Market with virtual impunity."[26] That was true in the sense that if any wholesaler had refused to pay tribute to the union, their fish would have gone unloaded and their business would have gone bankrupt. It does not mean that Romano gave orders about exactly how the market should be run. The whole point of paying off the mob was to keep them out of a company's day-to-day affairs. The relationship was parasitic, not controlling. The wholesalers were afraid of the mob because of the threat of physical violence and because they wanted to avoid problems with their operations that the union could easily have caused.

In 1984, the Federal Bureau of Investigation and the New York Police Department began operation Sea Probe, which tracked connections between racketeers at the market and other organized crime operations. In 1987, then U.S. Attorney Rudolph Giuliani filed a civil Racketeer Influenced and Corrupt Organizations Act (RICO) suit designed to put the Fulton Fish Market under receivership. The act had been designed to combat organized crime in the sense that it could be used against enterprises that planned and benefited from criminal activities. Unlike previous RICO cases, this one was based not on prior criminal convictions but on extensive connections between actors in the market and organized crime. It also drew a straight line from what Lanza did in the 1920s to what the Romanos did in the 1980s, noting how the mob's practices remained the same.[27] Giuliani built his case upon the idea that the mob controlled the union.[28] While that may have been true, it also made it possible to criticize Giuliani's efforts as simply political.

These charges against the union, the Genovese organized crime family, and the people who ran the union after the Romano brothers went to prison all failed. In 1989, U.S. District Court judge Thomas Griesa wrote:

> The proof offered by the Government has not been sufficient to support its essential claim, i.e., that the Genovese family currently controls, and in recent years has controlled, Local 359 and its principal officers . . .
>
> This is not to say that there is no evidence of the presence of organized crime, particularly Genovese family operatives, in the Fulton Fish Market. Vincent Romano [a third Romano brother who allegedly

succeeded Carmine and Peter after their convictions] is undoubtedly associated with the Genovese family and was regularly present in the Market over a period of many years.²⁹

Certainly the presence of the mob in the Fulton Fish Market did not benefit that institution—it became a serious problem for market governance—but it did not mean that organized crime controlled the market.³⁰

Before the RICO case ultimately failed, a consent decree signed in April 1988 allowed for the appointment of a market administrator and barred many of the people cited by the government from participating in any market activities, including all the Romanos. Although the government asked that the union be put under trusteeship, the judge in the case was unwilling to go that far. Frank W. Wohl began a four-year term as the court-appointed administrator overseeing the day-to-day operations of the Fulton Fish Market on May 3, 1988. In his midterm report in 1990, Wohl wrote that "racketeering activity persists in the Fulton Fish Market and . . . the current absence of legitimate regulatory control over the market creates an environment highly conducive to the continuation of such activity." He singled out the continuing monopoly of the unloading crews as the source of the mob's control.³¹

When the administrator's term expired, Mayor David Dinkins added two dozen new inspectors to the market, but this was not enough. When Rudy Giuliani became mayor in 1994, he proposed aggressive municipal regulation of the market as the solution to the decades-old problem. This was Local Law 50, which allowed the Department of Business Services to license all businesses operating at the market, set rules, and impose fines for violations of those rules. The rules also governed the firms that did the loading and unloading. The Department of Investigations received the ability to do background checks on everyone at the market and bar people with mob ties from working there. Shortly before the City Council passed the law in 1995, arsonists burned much of the Tin Building, destroying many records there. Citing that crime, the council overwhelmingly approved Giuliani's plan.³² The arsonists responsible for the fire were never arrested.³³

With the law passed by the council, the Giuliani administration broke up the monopoly that the unloading companies held over the wholesalers at Fulton Fish Market. It removed six unloading firms, raised rents on the stalls, and removed six wholesale fish firms whose personnel could not pass background checks. Instead of the mob-linked union firms, the new unloading contractor, Laro Maintenance Corporation (better known for cleaning offices around its home base on Long Island), did the same job for 20 percent less money than the corrupt unloaders had charged. The changeover was met with fierce resistance, which included a wildcat strike and sabotage on the lighting fixtures at the market, but it was ultimately successful.[34]

Without the mob-linked union, the wholesalers had no reason to pay protection money to the mob. This in turn lowered their costs of doing business, which in turn lowered prices for consumers. After Local Law 50 took hold, the volume of fish handled at the market increased to 183 million pounds per year (bucking a long trend) and the price of fish sold decreased by 2 percent even as the wholesale price of fish in the United States increased by 13 percent.[35] This was the argument that Giuliani had most often used when making the case to fight the mob before the public. Indeed, he and his allies often cited a "mob tax" on consumers imposed by organized crime.[36] Mayor Giuliani's general strategy against crime made a big deal about things that most people considered at worst minor annoyances, like broken windows. After all, not that many New Yorkers even bought fish that came through the market at this late date in its history.

"Just how strong a presence organized crime came to have in the Fulton Fish Market has always been a matter of contention," wrote Phillip Lopate in 2007.[37] To take sides in this dispute requires differentiating the companies that sold fish from the firms that loaded and unloaded them. The loading companies were essentially "granted monopolies by the mob," to use the phrase of one assistant U.S. attorney.[38] Determining the influence of the mafia over the fish wholesaling business is much harder because, as was the case earlier in the market's history, it is hard to disentangle one company from another. This makes it very difficult to know exactly how many wholesale firms were actually controlled by the mob and whether those that weren't submitted to the mafia's demands, willingly or unwillingly.

The people who unloaded the fish had the power to demand bribes to do their jobs. These same workers might take a few fish out of a box every once in a while. One of the unloaders told Mensch that this was "expense money."[39] That, along with the effect of any union on wages, might explain why rank-and-file market employees would have been happy with the mafia, but it says nothing about the wholesale fish dealers. "The mafia's presence and its culture of suspicion intimidated honest merchants and workers," later wrote Selwyn Raab, the organized crime reporter for the *New York Times*, "preventing them from cooperating with investigators or testifying."[40] In this sense, consent was immaterial because the corruption continued whether the wholesalers liked it or not.

Because organized criminals worked in the background collecting money from wholesalers, they had no need to control the day-to-day operation of the market. John Von Glahn, who eventually became the spokesman for the Fishery Council, told Joseph Mitchell, "Oh yes, I'd pass [Carmine's Bar and Grill] and see [Socks] Lanza sitting out there and I'd speak that is, I would nod and say, '"Good morning."' I never spoke to him by name. I somehow understood that you didn't do that. You know that guy sitting in a chair in the sun outside Carmine's: That's the Mafia guy."[41] If you controlled the fish market from a chair outside Carmine's, you were a king, not a field officer, benefiting from a system directly run by others. As long as the wholesalers kept paying off organized criminals, there was no need for those criminals to influence how the market ran on a daily basis.

Whether a wholesaler willingly went along with this kind of intimidation obviously depended upon a company's circumstances. There were at least three firms with direct family ties to the Genovese crime family. However, at least some firms were willing to fight against the mafia's control of the market by testifying in public before a New York City Council committee in 1992.[42] Other people had more idiosyncratic reactions.

"My neighbors all think I'm a gangster," complained one salesman in 1981.[43] If all that his neighbors knew about the fish market came from the local papers, it is easy to see how they might have reached that conclusion. Obviously, the history of the Fulton Fish Market involves much more than just the presence of organized crime. However, focusing almost exclusively upon that invariably implies a criminal influence over day-to-day operations that never existed. The mob controlled the market only in the sense that it had the ability to shut the place down by refusing to load or unload fish.

Because the mob also made no money by doing so, this happened only very rarely. There was, for example, a short strike for higher wages that shut down the fish market in 1982 and another to protest Giuliani's antimob policy in 1995, but these were rare exceptions to years of uninterrupted service.[44] Organized crime's influence over operations appears to have been minimal, except that many of those operations became more expensive because they required payoffs to the mob. Many of the fish that disappeared via "tapping" probably would have been wasted anyway because the cold storage facilities were both minimal and so behind the times.

If the mob had had no presence at the market, very little would have changed, at least in the short term. The wholesalers happily passed the costs of a corrupt labor peace down the supply chain. Changing the rules on the ground threatened their long-term stability. "As far as I'm concerned, the investigation was a waste of time and money," explained a "veteran wholesaler" to the *New York Daily News* after the conviction of Carmine Romano, "just a lot of publicity. There was no real problems (with organized crime); nobody is ever bothered. All we're looking for is to make a living."[45] The president of the Fulton Fishmongers Association famously claimed that it was more effective to pay the mob to prevent the theft of his product than wait for the police.[46] As long as both parties in these transactions benefited somehow, neither side objected.

However, the long-term effects of the mob's presence were more serious. By driving up costs in the Fulton Fish Market, the mob hastened the decentralization that made the place less and less important as the twentieth century passed. "It is said that the New Bedford fishing fleet used to put into Peck slip at New York quite regularly," explained a Connecticut paper in 1931, "but the unloading and unloaders' charges began to become too heavy a burden to make the practice profitable." They started landing their fish in Boston instead.[47] When the union struck to protest the Giuliani reforms in 1995, the city's best restaurants got their supplies from dealers who flew their fish into the region's airports and delivered it directly to their doors.[48] Other forces only accelerated this decentralization process over the decades, contributing to the willingness of every party to relocate to the Bronx in 2005.

Regardless of influence by the mafia, the culture of the Fulton Fish Market actually favored operating in the very short time windows that the lack of proper refrigeration forced upon dealers. As one salesman explained to Mensch:

> Today you're worth something, tomorrow you're nothing, it's the importance of *today*. Ya don't even have four hours to sell fish. In reality ya got two, three hours, tops . . . I get psyched. Excited. It all starts from the afternoon before, when I'm on the phone with my shippers. I really love the business. The day before, if I know I'm gonna have a lotta fish. I can't sleep sometimes when I have fish comin.' I'm like a tiger in a cage.
>
> In doing business I was taught to have self-confidence. If you don't have your own self-respect, if you don't have the confidence, the pride in what you're doin' an' enough respect, they'll pick up on it. They won't take you serious.[49]

That pride extended to selling quality fish, which meant fresh fish that had not been frozen.[50] No wonder that by the end of the market's time in Lower Manhattan, everybody who worked there was deeply annoyed by the fact that every writer who visited had to ask about the mafia. Most said that allegations of mob domination were overblown.[51]

It is impossible to blame just the mob for the failure to develop a cold storage system around the Fulton Fish Market because the culture of organized crime and the culture of the market fit together so well. The lack of cold storage was a problem long before organized crime arrived. Because the place was already wasteful, having some fish get "tapped" probably didn't seem like much of a disaster. Because the market had always depended upon long-term relationships based on trust, an organization that guaranteed labor peace must have been appealing. Because the easiest way to get a job at the market was to have someone in your family already working there, having an organization that wouldn't cross the color line running the least-skilled part of the operation helped reinforce tradition. That partly explains why when the forces of change finally reached the market, they all came from the outside.

Payments made by wholesalers to the mob acted, as the Giuliani administration suggested, like a tax on their operations. Even if the wholesalers passed those charges on to consumers of fish, the firms would have had an easier time surviving the lean years of the late twentieth century if they hadn't had to pay them. Wholesalers accepted the "tax" as a cost of doing business, but the effect on the efficiency of the market spurred its decline. The mafia also hindered the technological updating of the market's infrastructure because the lack of good refrigeration was its most important pressure point to establish power over operations.[52] At the same time, because of that pressure point, the mob was one of the most important parts of the coalition at the market that opposed moving.

The wholesalers could afford to pay the mafia because they had their own monopoly power over the fishermen and smaller dealers outside the New York market who wanted to sell there. "The fishermen," explained one writer in 1981,

> who rarely have deep cold storage facilities and must move their catch as soon as possible, argue that the Fulton Market has a stranglehold on their economy. Because the market will only handle fresh fish (though some shellfish does come in frozen) the fishermen must sell it at the offered price or risk being stuck with a perishable commodity. A catch of say, cod, might bring a thousand dollars one day to a Long Island fisherman, only half the very next day if there's a glut on the market. The money barely covers his operating expenses. Or, he might be faced

with having sold in a hurry only to find cod bringing four times as much later on.[53]

If the wholesalers had upgraded their market facilities, even by adding refrigeration that they themselves owned, they would have lost this particular bargaining chip. As long as there were still fish arriving, they could still make money in an inefficient system and pass some of their profit on to their "protection."

In other words, the most significant long-term impact of the mob upon the Fulton Fish Market wasn't the crime per se, it was the diversion of resources from the wholesalers that could have been used for the upkeep of all the facilities. "It used to be that everything would come into New York, which was the distribution point for Baltimore, Philadelphia, Connecticut and New Jersey, as well as New York," explained Frank Minio of Smitty's Fillet House in a 1981 article that was supposed to be about organized crime. "Then people went other places, building new facilities with other capabilities besides distribution—cutting, processing, freezing, designing new methods of preserving." Minio concluded, "Industrialization never really happened here. People were satisfied with the status quo. And now there's just no space to expand physically."[54] Industrialization, of course, would have meant the replacement of manpower by machines. Lack of industrialization and the presence of the mob in New York City contributed to the decision of many suppliers to develop new provisioning chains to take their fish. Shinnecock, on Long Island, for example, started its own cooperative during the 1980s in order to avoid organized crime in Lower Manhattan.[55]

If the mob hadn't been a presence at the Fulton Fish Market, the wholesale fish dealers would have made more money and perhaps moved to the Bronx sooner than they otherwise did, but the day-to-day operations of the market wouldn't have changed much. The neighborhood still would have changed, which in turn would have led to pressure for the market to relocate outside of Manhattan. The only thing the mafia influenced was the speed of that move. In the meantime, there was another force pushing Fulton Market out of its century-plus-old location, namely gentrification. Actually gentrification isn't the right way to describe the development of an area where so few people lived in the first place. Historic preservation, tourism, and the direction of a new museum also played a part in a whole series of changes that transformed the neighborhood around the Fulton Fish Market during the late twentieth century. All these developments culminated in the market's relocation in 2005.

13

A MUSEUM AND TWO SHOPPING MALLS

"The odor of this place is enough to nauseate any civilized person," wrote a reporter for the *New York Sun* referring to Fulton Market in 1869. "South street and the wharfs on either side of it are strewn with portions of the offal which is never thoroughly removed. The drainage is altogether insufficient; and, although the floors are frequently washed with water, the effluvia is disgusting in the extreme. The refuse and garbage of Fulton Market is placed in boxes or barrels outside the structure... and in hot weather emitting an odor the reverse of fragrant."[1] Later on, Joseph Mitchell described the smell of the entire neighborhood:

> The Fulton Market smell is a commingling of smells. I tried to take it apart. I could distinguish the reek of ancient fish and oyster houses, and the exhalations of the harbor. And I could distinguish the smell of tar, a smell that came from an attic on South Street, the net loft of a fishing-boat supply house, where trawler nets that have been dipped in tar vats are hung beside open windows to drain and dry. And I could distinguish the oakwood smell of smoke from the stack of a loft on Beekman Street in which final baddies are cured; the furnace of this loft burns white-oak and hickory shavings and saw dust. And tangled in these smells were still other smells—the acrid smoke from the stacks of the

row of coffee-roasting plants on Front Street, and the pungent smoke from the stack of the Purity Spice Mill on Dover Street, and the smell of rawhides from The Swamp, the tannery district, which adjoins the market on the North.[2]

The people who worked at the fish market were fully acclimated to these smells, but they were hardly inviting to affluent people who might want to move into what would eventually become a trendy Manhattan neighborhood.

After World War II, there were still comparatively few people living in the area around the Fulton Fish Market. The people entering the area from elsewhere came for Lower Manhattan's new office buildings. New residents came later, mostly in apartment buildings. Any kind of interest in the neighborhood made it harder to use the area surrounding the fish market for comparatively low-profit fish-related business. During this time, local authorities eagerly worked with different private developers to revitalize old neighborhoods throughout the city, especially in Manhattan. There was no single vision in the city's aim except change that promoted economic growth.[3] Around Fulton Fish Market, that change came from office buildings and tourism. A wholesale fish market that operated in the dead of night and early morning hours did not advance either of these priorities.

In 1953, the 1883 building that had been championed by Eugene Blackford (and had replaced the earlier retail markets going back to 1822) was replaced by a privately owned market building and parking garage.[4] Joe Cantalupo (the brother of Bob Cantalupo and a member of the mafia family that controlled the garbage contract at the Fulton Fish Market) was a close friend of Joseph Mitchell and another fish market history enthusiast. When the earlier building was demolished in 1950, they rescued the terra cotta statuary from the sides of the building before it fell.[5] That left two official market buildings: the 1907 wholesale building and the 1939 replacement for the secondary wholesale building that had fallen into the river. There were also operations out of storefronts all around the neighborhood. As late as 1978, the market area could still be described as a "sprawling, four-square block architectural anomaly of wooden sheds, tenement houses, a whiskey warehouse, stables, an unused theater, a gun shop, ship chandleries, and finally, a concrete and steel structure opened in 1939 that still comprises the physical plant we know as the Fulton Fish Market," but the rapidly rising value of real estate made those other operations increasingly unsustainable.[6]

Zoning changes in 1961 led to a spate of new construction near the market. Various developers built about a dozen skyscrapers or superblocks by 1967.[7] The problem for the fish market was not that these structures were built on land it had once used, but that the gradual expansion of downtown office space into the areas nearby drove up prices for land associated with the market, thereby making it harder for the owners of those buildings to resist selling out to developers. Developers bought up Schermerhorn Row, but local opposition prevented the demolition of that obviously historic structure. Peter Stanford, who would become the driving force behind the South Street Seaport Historic District and would later serve as the founding president of the South Street Seaport Museum, organized a petition campaign and encouraged residential tenants to squat in the building to make demolition more difficult. Later, the city intervened, offered nearby air rights to expand neighboring projects, and then preserved that now-famous block of early nineteenth-century federal structures.[8]

New arrivals came because there were new buildings in the area in which they could live or work. For that to happen, old buildings had disappeared. "Some more buildings have come down on Fulton Street," Joseph Mitchell reported in the notes of his walk from July 16, 1968, "including a colonial-style building that was put up by the Corn Exchange Bank for its Fulton Street branch and which I myself saw go up. I even remember the building it replaced . . . a dismaying thought. . . . They tore down an old building and put up something worse and now they've torn *it* down and they'll certainly put up something even worse. . . . The fish market was deserted."[9] Mitchell's complaint is familiar to New York City residents, who can often date their length of time living there by remembering the buildings that have come down, since so much of the city changes so often.[10] Yet as late as the mid-1970s, most of the buildings in the neighborhood were "shuttered and rotting from the inside,"

FIGURE 13.1 South Street Seaport, 1970s. The last two original Fulton Fish Market buildings, along the river, are visible in this shot. The initial signs of the redevelopment of the neighborhood are visible in the foreground.

Library of Congress, Washington, D.C.

remembered a local resident who first arrived in 1974.[11] No wonder the fish market developed a reputation for being in decline when these structures were the only things that most visitors saw. Even those who came late at night to see the market in operation would not have been impressed by the main facility's then decades-old infrastructure.

Two developments very close to the Fulton Fish Market spurred the transformation of the entire neighborhood into something new by the end of the twentieth century and beyond: the founding of the South Street Seaport Museum in 1967 and the development of the neighborhood by the Rouse Corporation, a Baltimore firm best known for its successful revitalization of the Faneuil Hall area in Boston. That started with a deal struck in 1979. Even though the museum supported the fish market's continued existence at its traditional location, its collaboration with developers and development undercut that support. More development increased rents. Businesses that made more money than wholesale fish dealers then bought up properties that the dealers had moved into earlier in the century, thereby changing the character of the neighborhood. The city and the state never deemed the actual fish market worthy of protection.

As a result, every new project that made the neighborhood more desirable made it harder for the fish market to stay a fish market. Back in 1966, in a telegram to Governor Nelson Rockefeller, his brother David explained his opposition to the bill creating the South Street Seaport this way:

> THE BILL, WELL-INTENTIONED BUT ENTIRE CITY BLOCK IN LOWER MANHATTAN, TOGETHER WITH FULTON FISH MARKET AND ADJACENT PIERS, LAND AND WATERFRONT PROPOSED FOR SOUTH STREET MARITIME MUSEUM, COULD BE USED BETTER FOR NATURAL EXPANSION OF FINANCIAL AND INSURANCE COMMUNITIES. CITY WOULD BENEFIT GREATLY FROM ADDITIONAL REAL ESTATE AND BUSINESS TAXES THAT COMMERCIAL STRUCTURE COULD PRODUCE.[12]

Rockefeller's position won out in the end. The people who wanted to use the land in the seaport area for the activities that brought the most money easily outbid other entities. Although intervention by the state saved some of the historic structures, it is telling that Schermerhorn Row had always been primarily a general commercial structure rather than directly controlled by the fish market. The fish market didn't attract enough tourists to justify its location in the eyes of developers, but they were willing to give a maritime museum a chance to entice more people into the neighborhood.

In 1966, State Senator Whitney Seymour, Jr., introduced a bill to create a maritime-oriented museum in Lower Manhattan. Around the same time, a former advertising executive named Peter Stanford organized the Friends of South Street, a group dedicated to telling the maritime history of New York City. Nelson Rockefeller signed the legislation despite the fierce opposition of his brother David, who ran Chase Manhattan Bank and apparently would have preferred that Wall Street firms and insurance companies simply gobbled up the neighborhood. The Friends of South Street received a charter from the State of New York Board of Regents in 1967 to create the South Street Seaport Museum. The bill had eliminated all public funding for the project, so the founders of the museum had to engage in a massive fundraising campaign.[13] The museum then formed a subsidiary called Seaport Holdings Inc., which used a donor's stock gift to buy up land in the seaport area. That effort collapsed in 1972 when the value of the gift dropped.[14]

The idea behind the museum was to preserve as much as possible of the old seaport, including restoring many of the early nineteenth-century houses in the neighborhood surrounding the market. Preserving the Fulton Fish Market was never central to this mission. On the day the plan for the museum was announced, the architectural consultant for the project credited the market for the preservation of the buildings in the surrounding neighborhood. "Possibly because it smells so bad people stayed away from here," William C. Shopsin told the *Times*.[15] During the 1990s, a resident who lived a half mile away remarked, "On hot days, the smell of fish was nauseating."[16] Everybody but the fishmongers stood to benefit if the source of that smell somehow disappeared. The museum's early leadership accepted the fact that the city would tear down the market as a vital part of their planning process. They cared much more about ships than fish.[17]

In its early years, the people who ran the South Street Seaport Museum knew that they couldn't save every building, so they tried to save as many as possible. "These buildings cannot be preserved unless they are adapted to the needs of today," explained an article in the museum's newsletter from 1972. "Only structures of extraordinary individual distinction can be preserved without a modern viable economic use."[18] That meant the early nineteenth-century structures of the port of New York, not the early twentieth-century structures of the Fulton Fish Market. Even if the author of that essay didn't mean to apply this standard to the market's structures, the museum's interest in the surrounding buildings increased speculation in neighboring real estate, which in turn made it harder for the wholesalers to hold on to their offices and work spaces. Development indirectly threatened the market by making it harder to resist calls for being moved.[19]

Originally, the people responsible for the restoration of the neighborhood picked 1851 as a target date, a time when Fulton Market was nowhere as near as important to the city as it would become just a few decades later.[20] The museum chose that date because its primary interest lay in the port neighborhood that preceded the fish market's best years. The name South Street Seaport, which illustrates that emphasis, dates from 1967, not the mid-nineteenth century.[21] It was a marketing slogan. By promoting the neighborhood in this way, both the museum and the developers who worked with it were creating an image that mostly excluded the fish market, even though it lasted a lot longer in the neighborhood than all the clipper ships that had docked there.

Within a few years, though, the museum's official position became much more supportive. By 1975, a *New York Daily News* reporter with very clear sympathies for the fish market could report, "The Seaport Museum itself, incidentally, is not all that eager to oust the Market."[22] While the museum took a strictly neutral stance with respect to the fish market's fate in public, the two institutions became friendlier over time. For the people promoting the city's heritage, the fishmongers became useful allies in their struggle with other forces who wanted to remake the neighborhood differently. By the 1990s, the museum led tours of the fish market during its early morning hours and ran school programs about fish.[23] Material related to the Fulton Fish Market remains an important subsection of the museum's collections and archives.

Unfortunately for both institutions, the developers interested in revitalizing the area cared more about profit than about preservation of any kind. They quickly started outbidding the museum for buildings close to the fish market that it wanted to buy for preservation purposes.[24] Buildings over a hundred years old in the area were a disincentive for investment because their presence made it more difficult to develop further.²⁵ As the museum's plans unfolded, they simply noted that the city had scheduled the fish market to move in 1977 and that its operations would remain unobstructed by the museum's development as long as it remained in the neighborhood. However, because much of the Seaport Museum's resources were in real estate, it actively encouraged the development of the area it controlled in order to improve its collection of ships. A development plan published in 1974 declared, "The Museum was founded as a central point for the collection of ships and marine artifacts, and the study and presentation of vanished ways of life that originally built New York City."²⁶ Fishing as a way of life had certainly diminished by 1974, but it hadn't vanished yet.

During this same period, the city pursued a strategy that sided with developers of the South Street Seaport over preservationists and the smaller fish dealers who wanted to keep operating in Lower Manhattan. "The seaport will bring people to the stores," explained the director of Lower Manhattan Planning and Development in 1972, "the stores will make the hotels and residential development feasible, and they in turn will make working in the area more attractive."²⁷ The redevelopment of the South Street Seaport was originally part of a much larger project called Manhattan Landing. Announced by David Rockefeller and Mayor John Lindsay in 1972, the plan was supposed to involve the redevelopment of all of Lower Manhattan into a twenty-four-hour community. It would have included six million square feet of new office space, 9,500 new apartments, a hotel, and a park. The city's financial crisis prevented this larger effort.²⁸ Nonetheless, the City of New York, the fishmongers, the Seaport Museum, and the developers held extensive talks following the announcement so that every side could benefit from this significant effort.

A $60 million deal among the Rouse Corporation, the Seaport Museum, and the New York State Urban Development Corporation in 1979 marked the beginning of a two-stage project that remade the neighborhood. This plan had been in the works since 1974, when the seaport's planner, Benjamin Thompson Associates, convinced Rouse to craft a proposal.[29] In Rouse's famous "festival marketplaces" in Baltimore and Boston, the museums involved served as landlords, taking rents from the retail stores in both malls to fund their operations and collecting. The city agreed to "respect the role played by the Fulton Fish Market as a vital institution in the Historic District" and "work with the Market and its constituent companies to define their needs and interests." When there were differences between the developers and the wholesalers, the city intended to "take steps . . . to resolve points of conflict."[30] These efforts at reconciliation largely failed. In Manhattan, the plan met considerable opposition from many interested parties, including the wholesale dealers at the Fulton Fish Market and many of their employees.[31]

In response, fish-related commerce became part of the plan for the first retail mall. "This structure will incorporate part of the existing Fulton Fish Market stalls along South Street, so that the sale—and smell—of fish will continue," explained an advertising booklet distributed with the *New York Times* in 1982.[32] When this mall was completed the next year, it also got fish decor, with replicas of large fish hung in the atrium over the food court.[33] This first stage, called The New Fulton Market, was built directly on top of the site of the first Fulton Market of 1822. (It replaced a private parking garage that had been built after the previous fish market-related building on that spot came down in 1953.) The museum deliberately invoked the history of the place in its promotional materials for the new retail facility. As one brochure explained, "The New Fulton Market deliberately brings a new vitality to the seaport. Its purpose is to recreate an historic activity—the marketplace."[34] It was a 32,000-square-foot shopping mall, with mostly the same kind of tenants that could be found in the suburbs at that time.[35] Calling it Fulton Market, a name that had fallen out of favor by 1983, suggests the efforts to downplay the history of the fish industry in the area. That was ironic because the fish market remained operational, whereas retail shopping had disappeared decades earlier.

The second stage, finished in 1985, involved another mall stretched out over the water called Pier 17.[36] Rather than having any ties to New York, its shops resembled the mix of stores in any upscale mall around the

FIGURE 13.2 The 1983 Fulton Market retail building was called "Fulton Market" to hearken back to the retail activity that once dominated that location, but its success proved very short-lived.

Wikimedia Commons

country. Worse still, it obstructed the view of the Brooklyn Bridge from many vantage points in the neighborhood. In 2008, the *Times* reported that Pier 17 "now feels like a run-down, sometimes eerily unpopulated attraction to which its neighbors seldom venture."[37] In 2004, Rouse, throwing in the towel, sold all its Manhattan retail holdings to a company from Chicago called General Growth Properties (GGP). In 2009, GGP spun off the malls and leases to the Howard Hughes Corporation.[38] Their 2013 decision to raze the mall on Pier 17, which had been damaged by Superstorm Sandy, and replace it with an entirely different one in the same location suggests that even the developers knew that retail redevelopment targeting the tourist trade had failed in this area.[39]

In their early years, both parts of the Seaport Project did very well. By 1985, the seaport led every other Rouse retail facility in sales per square foot. A 1991 poll in *Newsday* claimed that it was the most popular summer tourist destination in the entire city, beating out Central Park.⁴⁰ In the end, Rouse invested over $100 million (of a $350 million project) in the construction. It also took out an $89.5 million mortgage after the fact to finance further development of the property. Despite the early success, the company never made anywhere near enough money to cover its costs.⁴¹ Part of the problem was that the tenants in both malls kept changing because their revenues couldn't cover their rents, and the rents couldn't cover Rouse's costs.⁴² As early as the late 1980s, Rouse officials told Peter Stanford, "We're waltzing with a corpse." By 1998, the place was nearly empty. The city never made more than minimum rent on the property it retained throughout the area, as its revenue depended upon Rouse's revenue from the overall project.⁴³

Looking back, it is easy to see that the original plan for the seaport didn't work in the long run because the redeveloped neighborhood could never become all things to all people. "The irony is that the South Street Seaport has never been an unqualified moneymaker," wrote Phillip Lopate in 2007:

> Shopkeepers and restaurant owners have learned that its magnetic capacities are highly erratic and seasonal, so it can be jammed with tourists one day, deserted the next. The executives keep tinkering with the formula, but the seaport has so far failed to become an organic, functioning necessity in the daily life of the city. Native New Yorkers tend to give it a wide berth, as an ersatz theme park with the same franchise shopping mall clothing stores that have popped up every twenty blocks in Manhattan and few compelling lures. And now that the fish market is gone, it seems even less authentic.⁴⁴

This proved true for both structures from the original plan, the 1983 Fulton Market building and the 1985 mall on Pier 17. People moving back into the neighborhood must have provided some solace to the planners who examined their handiwork years later, but their critics from when the redevelopment plan was being debated were essentially correct. The character of the neighborhood changed even before the fish market moved, and it is unlikely to change back.

Eventually, these changes helped drive the Fulton Fish Market out of the district. Stores attracted tourists, and tourists didn't care about a market they couldn't see. Yet the tourists were not enough to sustain the businesses that catered to them. Moreover, this kind of development was supposed to put the South Street Seaport Museum on a stable footing, but instead both institutions struggled. The Fulton Fish Market was shrinking during its final decades in Manhattan, but at least there were firms within it that still made money. This is not to suggest that the market deserved to be the subject of its own potentially more successful museum or that it should never have had to move. The struggles over the neighborhood suggest that different elements of city life can replace a former industrial site like the Fulton Fish Market, and retail-based heritage tourism may not always be the ideal option. Perhaps it was for the best that the Fulton Fish Market was destined to be a wholesale facility in the Bronx rather than a mall on the site of a market that only operated on a retail basis during the early to mid-nineteenth century and presently sold very few fish.

Even during the years when the retail business boomed, there were plenty of signs that the redevelopment wouldn't help the fish market or related businesses around the South Street Seaport. In 1985, Bill McKibben (then of the *New Yorker*) visited the neighborhood around the Fulton Fish Market, took a tour, and met the painter Naima Rauam, a longtime resident. "We stood with her on the curb," he wrote, "looking west across Front Street at the remaining block of aging fillet houses, at the piles of fish stacked in front of them on the sidewalks, at the men with hooks (for grabbing gills) slung around their necks like stethoscopes. It was so picturesque it nearly seemed fake, except for the smell." Those fillet houses all dated from earlier in the century. "This area is being heavily renovated," Rauam told McKibben. "From a practical point of view it has to happen, but aesthetically it's a shame. So I'm painting fast to get it all down."[45] It's ironic that the event that spurred Rauam's desperation involved a pathbreaking collaboration on behalf of historic preservation.

The many projects that gradually surrounded the neighborhood clearly illustrate the huge difference in attitudes between the developers and the people at the fish market. Although the Rouse development did nothing to displace the Fulton Fish Market, years of construction inevitably inconvenienced the people who worked there. "We're not against the development as much as we're against their whole attitude towards us, which is 'Drop dead,'" one fish dealer told the *Times* right before the first phase of the development opened in 1983. "It's a one-way street with them."[46] Nevertheless, as Barbara Mensch wrote about market–mall relations around this same time:

> The officials presented maps, blueprints, and many pages filled with numbers, statistics, charts, and graphs. The representatives of the Rouse Corporation used words like "rezoning," "reconfiguring," "renovation," "restoration," "relocation," and "rejuvenation." Sitting at the other end of the shiny table, the fishmongers looked baffled, forlorn, uncomfortable, and angry. They were being outmaneuvered. The problem was clear, how could these disparate groups coexist on the same waterfront property now becoming more valuable than gold?[47]

The long-term trends in the neighborhood still threatened the market's very existence. Both the wholesale fish dealers and the shop owners who rented spaces in the neighborhood were being told that they weren't chic enough to stay after the redevelopment was completed. Similarly, pensioners were being kicked out of the local hotels so that the rooms could be turned into apartments.[48]

In other words, the possibility of relocating the Fulton Fish Market hung in the background of every building refurbishment and every new structure. Even though there were various handouts and nods to the importance of the fishing industry throughout the redevelopment process, the forces unleashed in the area worked at cross purposes with any effort to keep the fish market in its traditional home.[49] By 1983, various mayoral administrations had been doing their best to relocate the market for well over twenty years. If it was so important to the life of the neighborhood, why had so many administrations tried to move it? As more people moved in, new coalitions formed to shape that development and left the wholesale fish dealers out. After all, people had to get up in the middle of the night just to talk to them. How were the fishmongers supposed to play a role in the life of the neighborhood when they only worked there at the most inconvenient time imaginable? Yet if the Fulton Fish Market was a "vital institution" in a historic district, how could anyone expect it to relocate?

From a historic preservation standpoint, there was good news and bad news. The preservationists behind both the South Street Seaport and the South Street Seaport Museum managed to get an eleven-block area just south of the Brooklyn Bridge landmarked during the late 1960s.[50] That saved some buildings, but not others. For example, a whole block of early nineteenth-century early Greek revival buildings on South Street was one notable casualty of developments that occurred before landmark designation.[51] With only partial victory, the nature of the whole neighborhood still changed drastically over time. As *New York Times* architecture critic Paul Goldberger wrote when the Rouse project opened, "The seaport district has changed as dramatically, in a sense, as if all of its little old buildings had been demolished and replaced by skyscrapers. For the casual, disheveled air, the sense of funkiness that filled these blocks until a couple of years ago has now gone completely. The feeling the seaport blocks used to give, that of a poor group of 19th-century buildings holding on for dear life against the powerful march of the twentieth-century city and its skyscrapers, has now disappeared."[52] The 1982 advertising supplement in the *Times* noted that there were a total of five major construction projects within view of one another in the neighborhood at that time.[53] The neighborhood changed even more as time passed.

This drastic amount of change eventually overshadowed the nautical theme that had been the original justification for preserving the neighborhood buildings. This proved particularly true for the Rouse shopping malls. Although the company's contract required Rouse to "promote businesses featuring 'maritime and sea-related activities and products,'" only two such businesses met that description by 1998, and one of them was the fish market.[54] Superstorm Sandy severely damaged both the New Fulton Market retail facility and the mall on Pier 17. The new building that was supposed to invoke the historic activity known as shopping isn't even a mall anymore. It houses a theater, a restaurant, and a large open space inside where fashion shows are staged, like when the actress Sarah Jessica Parker wants to introduce new items to her store that anchors the far end of Schermerhorn Row. At least there are still a few copper plates of fish displayed on the outside of the building. (Actually, it's the same fish, duplicated over and over every few feet.) It's telling that invoking the sea, which was supposed to placate the traditionalists who wanted to see the fish market stay, was quickly forgotten when the development started struggling.

Other area institutions met their fates more quickly than the market itself did. In 1981, the Rouse Corporation renegotiated the leases with the restaurants within the boundaries of its development. Sweet's and Sloppy Louie's got new long-term rental agreements as well as renovations to their buildings.[55] Lena Morino was the wife of Amil Morino, Sloppy Louie's brother and copartner of that restaurant. "Since we have become part of the Seaport Museum, tourists from all over the world are visiting us," she told *The Villager* in 1981. "Lena is assured that Louie's will remain, despite rising real estate costs, 'because we're a landmark,' she says."[56] The Rouse Corporation forced Sloppy Louie's to close in 1998 by raising its rent. Sweet's couldn't recover from a Nor'easter that struck the area in 1992, but rising rent was one of the reasons its owners chose not to continue. That restaurant was replaced with an Ann Taylor store, selling upscale women's clothes.[57]

How long could the Fulton Fish Market itself resist? Indirect pressure from the transformation of the neighborhood might have forced it to relocate all by itself, but wasn't enough. More importantly, the market also faced a long political campaign by local government designed to make it relocate. The City of New York, which had been the fish market's landlord since its beginnings, became the most important force behind the move to the Bronx during its final years in Lower Manhattan. Both the presence of organized crime and the lack of modern refrigeration played significant roles in its eventual departure. So did the continued evolution of the demand for seafood in New York, as well as in the country at large.

14

RELOCATION

In 1854, an important New Yorker openly wondered whether "a more advantageous disposition may not be made of that valuable property by the removal of the Fish Market." In 1859, a similarly powerful person suggested moving the market uptown because the current of the East River was "not sufficiently strong to carry off the offal."[1] Fulton Market remained. In 1879, a representative of the estate that owned the land under Fulton Market invited the leaders of the Fulton Market Fishmongers Association uptown for a meeting. The fishmongers heard an offer to buy the building and operations and the market from them so that the land under it could be put to more profitable use than selling fish. The headline for the *New York World* article describing the meeting read, "Fulton Market Not to Go Uptown." Eugene Blackford, then the leader of the association, was asked, "Has your company acted on the proposition?" He responded, "We have not, nor have we ever considered it."[2] Fulton Market remained.

Other sources of New York's food supply moved out of the city over time. Farming takes up land that is too valuable to use for that purpose when fruits and vegetables can be grown in adjoining areas and transported fresh to the city within hours. As railroads expanded, those kinds of foods moved farther and farther away. There were significant problems in New York and other cities with urban slaughterhouses during the nineteenth century because of the smell and related issues. Starting in the 1880s, iced railway cars were the solution as most meat production concentrated in Chicago, and later expanded to other cities. For a time, there were advantages to having fish of all kinds to pass through Fulton Market on the way to locations across the United States, but developments in transportation and refrigeration made it less important for the rest of the country as the twentieth century progressed. A 1953 consultant's report for the City's Department of Public Works noted:

> The most important considerations in favor of the existing location [for the Fulton Fish Market] are the investment of the merchants in the market, the habit of the consumers developed in the course of many generations to go to lower Manhattan for their wholesale purchases, the intricate intramarket relationships between wholesalers, jobbers and specialists and the availability of established transportation and supply companies. None of the possible disadvantages of the present location of the Fulton Fish Market even remotely approaches in importance these ties based on long tradition and organic development.[3]

Fulton Market remained.

The modern effort to move the Fulton Fish Market dates from 1956, when David Rockefeller of Chase Manhattan Bank organized a group of businessmen to revitalize all of Lower Manhattan. Their original master plan, issued in 1958, "called for the removal of the Fulton Fish Market to create a major new landmass that . . . would permit the financial and insurance districts to expand to the east."[4] In January 1960, Rockefeller proposed the first iteration of what would become the World Trade Center. Its original proposed location was along the East River, very close to the market. "The whole area," explained an article in the *New York Times* that appeared shortly after the announcement, "is, by fairly common agreement, hopelessly outdated. It is unsanitary (although the streets are hosed) and inefficient, and those with vision of a future for Downtown long for the day when the whole works will go elsewhere."[5]

That "whole works" included the Fulton Fish Market and large sections of the historic neighborhood that surrounded it. As time passed, the plans for the World Trade Center on the East Side got more ambitious, ultimately coming to resemble Rockefeller Center—another complex of buildings all built at once, designed to replace an entire neighborhood.[6] One of the aims of this very broad project was to replace all the industrial buildings and houses in the area, which the planners considered "obsolete and deteriorated," with a vibrant modern residential community, thereby reversing the depopulation of eighty years before. The plan also included a 350-room hotel, office space, exhibition space, two separate skyscrapers, and a new home for the New York Stock Exchange.[7]

Rockefeller and his allies had called for the removal of the Fulton Fish Market years before they released the World Trade Center plan, but the possibility of building that huge complex made relocation of the market seem inevitable. In 1960, Gilbert Millstein of the *Times* wrote, "In a couple of years or so, it is now pretty well settled by the people who decide these things, Washington Market and the Fulton Fish Market, two of the landmarks by which New York has best identified itself, will be pulled down and set up somewhere else, probably in the Bronx." Instead, the World Trade Center went in much closer to the west side of Manhattan because New Jersey played an essential role in the Port Authority and the governor would not sign off on a major project that was too far away from his state to benefit his constituents.[8] The Fulton Fish Market remained.

Besides the push from developers and others who wanted to use the land and facilities that the Fulton Fish Market occupied, there was also a pull from the City of New York, which developed a plan to centralize food distribution of all kinds in the Bronx during the early 1950s. In 1955, the head of the city's Department of Markets, Anthony Maciorelli, told the City Planning Commission that commercial fishermen were bypassing the Fulton Fish Market and landing their catches in Boston, Gloucester, and New Bedford because those ports had municipal support, which translated into much better facilities.[9] As part of the overall campaign to build the World Trade Center, the city condemned Washington Market in 1960. Older than the Fulton Fish Market, it consisted of dealers who sold wholesale produce on the Hudson River side of Lower Manhattan. The wholesale portion of that market lasted until 1962, when it moved to a new modern facility at Hunts Point in the Bronx. The city's overall plan was to create "modern, self-sustaining facilities for the receipt, sale and distribution of all food commodities." This included fish.[10] The city's meat processors and dealers began to leave Manhattan and move to Hunts Point around this same time.[11]

In 1963, Mayor Robert Wagner began an extended search process to find a new site for the market.[12] Later that year, the city released a consultant's report recommending that the Fulton Fish Market be relocated. It concluded:

> Located astride South Street, a major north-south artery with thousands of vehicle movements daily, it is difficult for buyers to come and go freely. The rundown, unsanitary and inadequate market buildings are hardly attractive to discriminating buyers and certainly do not stimulate increased sales. Furthermore, our studies indicate that little, if anything, can be done to improve conditions at this location. It is in the best interest of the City and market users to relocate rather than rehabilitate.

The city's Department of Markets endorsed the plan.[13] Wagner eventually settled on Hunts Point as the place for the new fish market. In the 1966 Lower Manhattan Plan, the inspiration for the remaking of all of downtown New York City, the Fulton Fish Market is labeled on one map as "Obsolete Market."[14] In 1967, wholesalers at the market were told by "responsible city officials" that "they were going to have to move, like it or not, and if they didn't like it, one dealer recalls being told, they and their fish would be greeted one morning by bulldozers." One of the advantages of the new Hunts Point location was a substantial increase in freezing capacity, a persistent problem at the original market.[15] Yet the Fulton Fish Market remained.

The city approved a lease for the new market in the Bronx in 1969.[16] A new cooperative of fishmongers formed to take possession of the facilities, an architect drew up blueprints, plans were drawn for sewer lines and such, but the building remained unbuilt. When the first architectural firm hired to design it failed, another firm drew up plans.[17] "The word on the waterfront," the *Times* reported in 1973, "is that the city has finalized plans for an ultra-modern fish-handling installation (if those are the right words) at Hunts Point in the far reaches of the East Bronx." In 1974, Mayor Abraham Beame officially announced, at the fish market itself, the demolition of the old market buildings and the beginning of construction of the new market in the Bronx. When one of the fish market workers complained that this would lead to "guaranteed muggings," Beame replied, "A big guy like you is worried about getting mugged?" The city thought the moving date would be July 1976.[18] The Fulton Fish Market remained. In 1984, the city signed a new lease with the wholesalers that could have guaranteed the market's staying put all the way through today.[19]

In the short term, the workers made arguments against moving the fish market that seem small and self-interested in retrospect. These included the difficulty of getting to Hunts Point and the possibility of job cuts due to automation. Crime—in other words, the possibility of getting mugged in Hunts Point early in the morning—was another argument against the move, even though many of the people who worked at the market were indeed large, strong men.[20] There were also self-interested arguments made by the people who owned businesses in the neighborhood around the market in Lower Manhattan. The most likely explanation for the failure of the city to move the market during the mid-1970s was its legendary budgetary problems, since there wasn't enough money available to build a new modern facility in Hunts Point.[21]

Nevertheless, the back and forth among different interests in the city reveals a great deal about how various groups exercised their power. Much of the pressure brought to bear on the wholesalers originated from developers who wanted to build in the neighborhood and saw the presence of the fish market as an impediment to their plans. As an official in the Office of the Lower Manhattan Development explained in 1975, "Presumably the owners and purchasers of property in the area are aware that the Fulton Fish Market is scheduled to move to Hunts Point, and that the blocks have the potential for a much higher return on investment if they are developed as office buildings."[22] While this was certainly true, the question of whether to move the market involved a lot more than just new development.

If anything, the back and forth between parties that actually worked in the market is even more important for understanding its history at this time. Most notably, the chairman of the Committee to Save the Fulton Fish Market was Carmine Romano—the same Carmine Romano who was president of Local 359 of the United Seafood Workers Union and who the federal government would later prove to be the representative of the Genovese crime family in the market. "We've offered to match them with $10 million of our own money to renovate and improve the market as it is right here, but they don't want to talk about it," Romano told a reporter from the *New York Daily News* in 1975, a point in time when the move to Hunts Point appeared to be imminent. "If they move us, they're going to put about 200 of my men out of work and put a lot of the smaller wholesalers out of business."[23]

Romano's self-interest in making this argument should be obvious. Move the market, automate the loading and unloading process, and offer better cold storage facilities, and his organization's choke hold over the wholesale fish dealers would disappear. The mafia's opposition to better cold storage facilities dated back to the 1930s. After the Seabury investigation finished in 1931, Joseph Lanza specifically targeted the Brooklyn Bridge Cold Storage Company, the largest supplier of refrigerated space near the market. He demanded of the firm's owner that all his employees join the union; otherwise, nobody at the market would move fish into his freezers. After holding out for three months, during which the firm lost 90 percent of its business, the owner agreed to sign a contract with Lanza's union.[24]

Romano and the rest of the opposition still made arguments against the move that had a strong basis in reality. "Of the 60 wholesalers," Romano explained in 1975, "there are only about three or four who are in favor of moving. . . . They're all big outfits, and they can afford to move. But it's the little guy that's going to get wiped out."[25] Large firms that could afford higher rents hoped that the increased efficiency would lead to greater volume and greater revenues. Joseph Mitchell, no friend of the effort to move the market to the Bronx, described the reasoning of those in favor as "room to move around in: freezer space, parking space, mainly loading space, display space."[26] These larger firms had access to fish from farther away and interests in wholesaling operations in other cities. Smaller firms opposed it because they were afraid they wouldn't survive the transition.[27]

"If we ever do finally move," Romano told the same reporter, "we'll lose all those guys from Jersey and Philly and Connecticut. They'll find something closer to home rather than drive all the way up here."[28] Most of the fish that came to the Fulton Fish Market already landed outside of New York City, so the comparatively local product was hardly missed. The new seafood provisioning chains, based on modern freezing techniques and rapid point-to-point transportation, had so severely cut into the advantage the market once held that whatever monopoly power it had had completely disappeared. Moreover, there was no particular benefit from processing or coordinating sales at a single location when modern communications made it possible to keep track of inventory even far away from wherever that inventory was located. A neighborhood that had once been largely devoted to receiving, selling, and processing fish had shrunk to two buildings, and those two buildings had become hopelessly obsolete.

From a long-term perspective, the geographical advantage of the Fulton Fish Market disappeared when fish stopped arriving there by water. Before that, the market had to be situated along the water. When they arrived in New York by train or truck, it no longer mattered where in the city the fish market happened to be. In fact, with the arrival of modern refrigeration and freezing, the largest fish market in America could have moved to Connecticut, or South Carolina for that matter, and still would have been able to serve the same fish to the same people anywhere in the United States because the supply chain depended as much upon trucks and trains as on proximity to the docks of a large number of fishing boats. Luckily for high-end chefs, with modern communications in place, there was nothing to stop wholesalers located outside New York

City from entering international markets too. With the value of the land under Fulton Market increasing, the need for its operations to remain there became far less urgent.

Besides the traffic and facilities problems, the 1963 consultant's report cited competition from frozen fish as a major reason to move the market. Even though New Yorkers were eating as much fish as ever, they wrote, the amount of fish going through the market was steadily declining. The reason was "less fresh, whole fish being sold for consumption. Packaged frozen fish and fish products in jars are shipped direct from the points of production, thus by-passing the Fulton Fish Market."[29] Cheaper, more abundant fresh fish meant more refrigerated fish, but lack of freezer capacity lasted throughout the history of the old market. When Corby Kummer visited in 1995, a buyer for a retail fish store told him, "'You wouldn't believe the storage conditions here.' . . . For years, vendors have had to rely on ice to display the fish, refrigerated display cases being a luxury reserved for retail merchants."[30] At first, less fish going through the market meant less fish left over at the end of the day, which meant less need for freezer capacity.[31] Ultimately, though, the forces that wanted to move the market used this deficiency against the wholesalers who wanted to stay put.

Lack of refrigeration and freezing capacity was always the reason that city officials most often cited when making the case to move. After all, there were health laws to back them up, and upgrading the existing facilities was just too expensive.[32] Then, new federal regulations in 2001 forbade the sale of fresh fish outdoors and required access to mechanical refrigeration at the point of sale. "Fishmongers at the Fulton Fish Market continue a long tradition of selling from open air stalls, where fish is kept on ice," reported the *Times* that year.[33] By the end of its time, conditions at the market were filthy. Pigeons defecated on fish left out in the open.[34] An outdoor fish market may have been all right in earlier decades, but the twenty-first century had arrived, bringing new standards with it.

FIGURE 14.1 Beekman Dock Icehouse, 1982. The few refrigeration-related facilities at the Fulton Fish Market in its last decades were both old and insufficient to meet the needs of a modern facility.

Photo by Barbara Mensch, in *A Falling Off Place: A Unique Story of Lower Manhattan*, Empire State Editions (New York: Fordham University Press, forthcoming)

As the capacity of trawlers increased with their overall size, deep-freezing technology got both colder and more efficient. Many of those innovations occurred during the 1960s. In 1981, the Japanese developed a small deep freezer for merchants.[35] The cumulative result was the first complete cold chain for fish, all the way from the point when they were caught to the retailers that sold them. When fish could be perfectly preserved from boat to plate, the Fulton Fish Market lost its chief remaining advantage over provisioning chains that never passed through New York City. The fact that many of the remaining wholesalers during the 1990s cut corners when it came to keeping their products fresh only hastened the departure from Manhattan.[36] The market became less and less important for obtaining fish, even in New York City, as it became easier and easier to obtain quality fish from provisioning chains that never went through Lower Manhattan.

The narrowing difference in quality between fresh and frozen fish meant that the two alternatives became hard to distinguish. Flash-frozen fish is still not quite the same as fresh, but it was close enough for most consumers in the late twentieth century. No lesser authority than James Beard, writing in *James Beard's Fish Cookery* from 1954, told his massive audience of home cooks, "The frozen food companies now produce a wide variety of frozen fish, and their selections are often excellent buys. If you live far from the fresh supply, or if you have your heart set on a fish that is not at the height of its season, the frozen product can solve your problem with little or no sacrifice in flavor or texture."[37] Home cooks who believed him would have been unlikely to pay for fresh fish from the Fulton Fish Market. The need to cook that fish almost immediately upon arriving at home would have made consumers less likely to pay extra in the first place.[38] Between the 1980s and early 2000s, the share of fish consumed in America traded by fish markets dropped from 65 percent to 11 percent. At the same time, the share of fish sold at supermarkets grew from 16 percent to 86 percent.[39] If it were served in a strong enough sauce, even the most fussy palettes would have had difficulty recognizing the difference.

One third of the fish consumed in America is ordered while eating out.[40] In 1999, Anthony Bourdain wrote a legendary essay in the *New Yorker* that explained when not to order fish in New York restaurants. It launched his entire postcooking career. "The fish specialty is reasonably priced, and the place got two stars in the *Times*. Why not go for it? If you like four-day-old fish, be my guest." He then explained the system that made this situation possible:

> The chef orders his seafood for the weekend on Thursday night. It arrives on Friday morning. He's hoping to sell the bulk of it on Friday

and Saturday nights, when he knows that the restaurant will be busy, and he'd like to run out of the last few orders by Sunday evening. Many fish purveyors don't deliver on Saturday, so the chances are that the Monday-night tuna you want has been kicking around in the kitchen since Friday morning, under God knows what conditions.... Even if the chef has ordered just the right amount of tuna for the weekend, and has had to reorder it for a Monday delivery, the only safeguard against the seafood supplier's off-loading junk is the presence of a vigilant chef who can make sure that the delivery is fresh from Sunday night's market.[41]

The fact that any wholesaler could contemplate offloading four-day-old fish, or that a restaurant with two stars in the *Times* might serve it, demonstrates that New York restaurantgoers didn't have much appreciation for fresh fish at this time. This is pretty damning.

With very little freezing capacity, the Fulton Fish Market had justified its location on providing a product that the public couldn't fully appreciate. Bourdain later claimed his advice was just a New York thing, which meant that the market wasn't even particularly good at training the chefs it sold to either.[42] High-end chefs helped coax the Fulton Fish Market into upping the quality of its offerings, but its relative lack of refrigeration facilities had already undercut its business for decades by that point. The need for speed in selling, buying, and cooking the fish also raised the price. Chefs who cared about both quality and price inevitably began to purchase their seafood elsewhere.

Starting in the 1970s, fish buyers in New York City could go to sources outside the city and still get a fresh product. For example, when the newly revamped Grand Central Oyster Bar reopened in 1974, it got its lobsters straight from Maine, its oysters direct from Virginia and Long Island, and even some of its fresh fish from a supplier in New Bedford, Massachusetts. The restaurant still had a buyer who went to the fish market every day, but the supply sources were deliberately diversified.[43] In response to this new reality, the larger Fulton Fish Market wholesalers had interests in other supply hubs in cities across the country. That way, if Bostonians or Philadelphians wanted to eat southern fish like gray mullet or red snapper, they had no trouble bypassing New York City entirely. New York companies could simply arrange to fly it to them directly.[44] Anyone outside of coastal areas could get a national seafood broker to fly their desired fish to them directly too.[45] In the mid-1980s, the city built Fishport in Brooklyn. It was a state-of-the-art landing and processing site for fishing boats that still wanted to directly unload their catches for the New York market. It failed quickly.[46] Only New Yorkers had to suffer an indirect price increase imposed upon them by their wholesale market's inefficiency.

The city had to make a big investment to build the new market, and as early as 1974 was afraid that the fish business would completely dry up and it would be left with a facility that couldn't be converted to other uses.[47] The sheer number of new provisioning chains that bypassed the Fulton Fish Market became the final reason for the move. "A.&P. used to be down here, Food Fair used to be down here," explained Robert Kirby, a salesman for the firm of Joseph H. Carter who had been at the market for forty years by the time the move to Hunts Point finally happened. "Now they've cut the market out altogether."[48] Because most frozen fish could never go through the market, the catch from the first factory trawlers may have landed elsewhere, but that cod and other whitefish was more likely headed into fish sticks than to the plates of New York's fine restaurants anyway. Fresh fish went through the market, but a lot of expensive product was ultimately wasted because it could only stay there so long. Over time, these disadvantages chipped away at the market's reason for continuing to exist at its original location.

After years of effort to move the market, predictions became self-fulfilling prophecies. In other words, people predicted for so long that the market would move that they had to be right eventually. During the ongoing struggle over the location of the market, the city often got distracted from its long-running efforts to force the wholesalers to move, but the same forces that decentralized fishing distribution—improvements in refrigeration and transportation—made the advantages of keeping the fish market in Manhattan increasingly irrelevant. Its original location had to be near the water because that's how all of the fish arrived at the

FIGURE 14.2 Night Before Moving Day, 2005.

Photo by Barbara Mensch will appear in *A Falling Off Place: A Unique Story of Lower Manhattan,* Empire State Editions (New York: Fordham University Press, forthcoming)

beginning. Not just fish stopped arriving that way. Over time, trains and trucks replaced fishing boats, and other goods that did arrive by water tended to be delivered in the New York area at the port of Elizabeth, New Jersey. Most of the docks around Manhattan fell into disrepair.

The last boat to regularly dock at the Fulton Fish Market was the *F. V. Felicia*. Based in New Bedford, Massachusetts, it rarely went there except for repairs. It was 91 feet long, it had a wooden hull and a diesel engine, and it specialized in catching scallops. The boat was too old to survive a trip to Georges Bank, so it tended to visit the shoals off the coast of New Jersey and then unload its cargo at Pier 18 near the market.[49] Its last trip through the harbor and up the East River was in 1979.[50] That was the last direct link between the market and the water. When every fish or piece of seafood landed elsewhere, there was no longer any particular reason that the city's fish market had to be located along a shoreline, let alone one with real estate as expensive as the shoreline of Manhattan. This left an opening for people who wanted to change the way that the market worked to start playing hardball.

"It would be better for the city not to have the Fulton Fish Market if it's going to be operated the way it has been for the last 40, 50, 60 years," declared Rudy Giuliani in 1995, while imposing his antimafia reforms on the mostly unwilling workers there.[51] Of course, nobody wanted New York to go without a fish market. The idea was to use the presence of the mob as a cudgel against the wholesalers who didn't want to move. Giuliani successfully turned moving to the Bronx into a moral issue, which in turn helped increase public support for the decision. The Giuliani administration revealed its plan to move the Fulton Fish Market in February 2001. "We wouldn't be here doing this if we didn't clean up the Fulton Fish Market that presently exists," he explained, indirectly admitting that the presence of the mob had helped to keep the market in Lower Manhattan up to that point.[52]

There was a temporary move to a Bronx parking lot in the wake of the September 11 attacks, as the market was too close to the destruction of the World Trade Center to be safe for the first month and a half or so, but groundbreaking for the new facility in Hunts Point occurred in November 2001.[53] The initial projection was that the move would happen by the end of 2002.[54] However, because the site of the New Fulton Fish Market had been a manufactured gas plant, there was severe environmental contamination that had to be cleaned up first. Those costs, which hurt the city more in the wake of the 9/11 attack, delayed the move.[55] So did an extended legal battle over whether the firm overseeing the market to keep the mob out would go with it.

The move to the Bronx was rumored to be happening all through 2005. It was delayed twice that year, the first time because of construction modifications, the second time because of a lawsuit filed by Laro

Services Systems, the company that the Giuliani administration had hired to do the unloading at the old market after the passage of its anti-mob reforms. Laro argued that mafia-connected individuals were being allowed back into the unloading business at the new market. Laro Services managed to get a judge to issue an injunction against unloading fish at the new place without them, which caused a delay from September to November, when that order was lifted.[56] The city settled the lawsuit by allowing Laro to maintain its exclusive contract for three years.[57]

The last day at the old Fulton Fish Market was November 11, 2005. The market moved to the Bronx over a single weekend. It had been the last major fish market in the entire Western world to remain at its original location until then.[58] The first day at the New Fulton Fish Market was November 14, 2005. Mayor Michael Bloomberg, responding to a question about "character loss" at the opening, said, "Well, I think it's true if you like crowded streets and unsanitary conditions and dangerous working facilities, then you'll miss the last thing. This is all to the good.... Things change. The world changes, and we've got to keep up with giving people what they really need to live and that is jobs, and good food, and safe conditions and economic development and this facility really does all that."[59] Tellingly, most of those advantages went directly to consumers rather than the people who worked in and around the market.

The move to the Bronx finally happened for a variety of reasons. These included economic development in the neighborhood surrounding the market, changes in the way that seafood was distributed throughout the United States, and changes in the kinds of seafood that American consumers wanted to eat. But it is worth noting how successful the wholesale fish dealers in Lower Manhattan were at staying put for so long. Different firms came and went, but there were always enough wholesalers in the Fulton Fish Market who adjusted to remain profitable. Even with poor refrigeration facilities, their experience at buying and selling fish enabled some of them to still prosper even as business conditions became increasingly difficult. Many of those difficulties continued after the move to Hunts Point, while in Lower Manhattan, the remnants of the old Fulton Fish Market began to disappear.

CONCLUSION

After Relocation

In 2007, Nina Lalli of the *Village Voice* visited the New Fulton Fish Market in the Bronx and asked the people who worked there how they felt about the change. She explained, "Most buyers and sellers I talked to agree that the new facility is infinitely better for the fish than the old one at South Street Seaport, where the product was exposed to the elements—which in summertime threatened freshness constantly. The new space is enclosed, brightly lit, and just cold enough to see your breath."[1] But they still missed the old digs.

In 2010, the radio personality Ben Sargent visited the New Fulton Fish Market in order to contrast the new market with the old. "The days of tourists, people coming out of the clubs and talking to the fishermen . . . You know, the fun stuff is over," he said. "This is pretty much just where they come to make money." Like other visitors, Sargent found lots of mixed feelings about the new setting. The people he interviewed recognized that the superior facilities were good for business, but they missed the ambience of the old place. "In here it's like prison," one fishmonger told Sargent. "They swipe your card, you come to jail."[2] The building is isolated, imposing, and prisonlike, which seems appropriate for a place that sits across the East River from Rikers Island.[3]

To appreciate the difference between the old and the new market, it should be recognized that there wasn't all that much left of the fish

CON.1 The New Fulton Fish Market is located in the Hunts Point neighborhood of the Bronx. Because it is a modern wholesale facility, it has proven much harder to romanticize than its predecessor.

Wikimedia Commons

market in its final years in Lower Manhattan. Although there were still two market buildings, the extent of the business done there had been shrinking for decades. A lot of the sales that still happened simply were conducted out in the street. *The Daily News* described the scene in December 1995:

> Wobbly tables filled with ice jut out from the dozens of stalls along South St. On top lie masses of fish in shades of pink, orange, gray and brown.
> The cobblestone streets are filled with melting ice and water. Wooden packing crates lie strewn about.
> Every few feet, fires blaze from oil-drum trash cans as workers burn the used crates to help fight off the cold.[4]

During the nineteenth century, a lot of what happened at the old fish market happened indoors. During its final decades, almost all of the business occurred outdoors. So little physical infrastructure even necessary made it much easier to move the place. It also meant that

CON.2 View of the market from FDR Drive, 1980. In its final years, much of the business at the Fulton Fish Market occurred out on the street rather than inside the buildings.

Photo by Barbara Mensch will appear in *A Falling Off Place: A Unique Story of Lower Manhattan*, Empire State Editions (New York: Fordham University Press, forthcoming)

whatever new fish market the city built would be a huge improvement in terms of the efficiency and safety of handling seafood.

New York's new municipal wholesale fish market cost the city $86 million to build and is as long as the Empire State Building is high. At 400,000 square feet, it is still the second biggest fish market in the world, after the Tsukiji Fish Market in Tokyo. Besides modern refrigeration, the new facility has more loading space and on-site offices. It is operated as a cooperative made up of the firms that rent stalls there. Running counter to the inefficient reputation of the old market, the cooperative notes that "Each tenant has a responsibility to all tenants and to the Cooperative, in contributing to sanitation, security, and the safe and smooth flow of traffic in the market, and to its efficient operation." Among its many rules, the cooperative prohibits "Annoying or harassing of any person." Despite all these improvements over the old market, the new market only handles one third of the seafood consumed in the city.[5]

The decentralization represented by that statistic has forced the Fulton Fish Market's wholesalers to care more about the quality of their product than they once did. The high-quality refrigeration available at the New Fulton Fish Market is good for the fish and therefore presumably good for business. "We used to unload product in the streets," one fishmonger noted when a reported for the *New York Times* visited him in 2006. "Right now it's probably 85 degrees," another fishmonger told him. "My fish would be rotting in the street. It cannot withstand the heat temperature, so right no I'm a happy guy."[6] Sandy Ingber, the buyer for the Grand Central Oyster Bar, told the same *Times* reporter that he preferred the market's new space because of both the superior refrigeration—a permanent 40-degree chill throughout the facility that was good for the fish—and its cleanliness. "You can lick off the floor there almost," he said.[7]

Back in Lower Manhattan, a fight over what to do with the vacated market buildings ensued. A spokesman for the city's Economic Development Corporation told the *Times* in reference to the old market, shortly after it closed, "There was nothing remotely that could be of value, or serve as a memento. It was all really trash."[8] The Howard Hughes Corporation disassembled the Tin Building, one of the two remaining structures from the market inside the historic district, in 2018 and began reassembling it two years later, about thirty-two feet east of its original location. This was done to make the building less susceptible to environmental damage, but after the fire in 1995, there was very little of the original Tin Building left. In reality, it is a re-creation of the old building with a few historic elements like the columns, beams, and trusses.[9] At this writing, it is being reconfigured as an upscale market food hall.[10] The city denied the New Market Building (which Fiorello La Guardia opened in 1939) landmark status in 2013 demolished it in late 2021 after it had become a homeless encampment and a murder occurred there.[11] In short, none of the buildings that housed the operation of the Fulton Fish Market at any point in its history remains intact.

You can still see lettering spelling out the names of firms that occupied buildings in the neighborhood outside the main Fulton Fish Market painted on the bricks. Some of the associated structures from that earlier time are still standing, like Schermerhorn Row and one of the houses on Peck Slip that had been used as part of the freshwater fish market. There are still metal stalls lining South Street that were once the nighttime homes of fish wholesalers selling their product to the buyers who flocked under FDR Drive to see what was available. Many of the restaurants and bars in the neighborhood retain nautical themes. For example, Dorlan's Tavern (although spelled incorrectly) deliberately invokes the name of the two famous oyster houses that all by themselves drew New Yorkers to Fulton Market during the late nineteenth century. Despite these remnants, in the end the critics of the proposed move to the Bronx were essentially correct. The neighborhood around the Fulton Fish Market changed irrevocably after the market left, and almost all of the unique aspects of the place have completely disappeared.

The idea that there was something quintessentially New York City about the old fish market also dates from this earlier era. It was one of the reasons Joseph Mitchell frequented the place, and it only grew once the Fulton Fish Market's location became threatened. At the height of the effort to move the market during the 1970s, Dick Ryan of the *Daily News* focused not on the location, but on the people who worked in and around the market as the thing that made it special:

> But those who have been going down to Fulton Market for decades might argue that the beauty of traditional New York, especially at the first shimmering shafts of dawn along South St., has nothing to do with the inside of a museum.... It is Ernie Kroog, wiping his hands on his apron before peeling off a couple of strips of bacon for a trucker girding for the long haul back to New England. It is the smell and commotion before the sun has bobbed above the skyline. The fish themselves, crates and trucks of them—long, flat, squat, shimmering rainbows, exquisite eyes, beheaded, buried in a shell, scaly, exotic. The grumpy panorama of men going out to load bins and drive trucks and sell their

day's supply and go back to bed again. It is a microcosm of the glory of the working man. It is a grimy, sweaty ballet of New York across a century and a half.[12]

This case for not moving the market implies the same answers that the workers gave once the market actually moved: it would be (and later was) good for the fish, but bad for the people. To Ryan, Joseph Mitchell, and the defenders of the old location, the people in the market were more important than the fish.

Of course, the Bronx is still a part of New York City. However, the complex created there for the benefit of fish has proven much harder for anyone to romanticize. From the old fish market you could see New York landmarks like the Brooklyn Bridge, the Statue of Liberty, and (while it stood) the World Trade Center. The Fulton Fish Market predated all those other sites. To move it meant altering the landscape, so it naturally attracted defenders who wanted to keep New York City from changing quite as much as it always does. If every person who worked in and around the market had moved to Hunts Point in 1975, advocates for the old market would not have been satisfied. The notion of working-class people doing industrial work in a borough that was becoming increasingly affluent and white collar fired the resistance to moving it.

In ancient Greece, the marketplace was the center of daily life. The body politic congregated there to interact, make collective decisions, and conduct commerce. Fulton Market bore some resemblance to this gathering place during its early history, but its operations became less public as it evolved into a wholesale market. Nevertheless, it was public enough that anyone who wanted to see how their fish got to them could still get up early and visit. The fish business persisted long after other blue-collar industries had completely abandoned Manhattan. Today, without a subway stop anywhere near it, average New Yorkers would have difficulty getting to any of the city's wholesale markets in the South Bronx. Moreover, because of improvements in refrigeration and transportation, wholesale markets aren't even necessary for restaurants or groceries to operate in the city anymore. The very long provisioning chains that the Fulton Fish Market pioneered helped eliminate what had once been the close connection between New York City and its surrounding waters. These days, it is very easy to forget that Manhattan is an island.

Fish from local waters once defined local food culture across the United States. Oysters from New York waters, crabs from the Chesapeake Bay, salmon from the Pacific Northwest—when these local favorites became international favorites, supply became a problem. The commodification of fish, allowing it to be sourced from anywhere in the world with a minimal effect on taste, has obscured the environmental damage that both the fish market and the market for fish caused. The depletion of local species happened as the volume of the fish going through the market dropped. As fishing techniques became even more efficient in the mid-twentieth century, the trawlers fishing off the shores of the United States began to ship loads not just to American markets but elsewhere too. With everyone's catches dropping precipitously, the federal government finally acted to conserve America's fisheries for itself. The Magnuson-Stevens Fisheries Act of 1976 allowed Congress to regulate waters under federal jurisdiction for the first time and created regional councils to establish management plans for depleted fisheries. The Sustainable Fisheries Act of 1996, a substantial revision to that law, allows the federal government to survey fisheries and close them entirely so that a rebuilding plan can be established if the amount of the stock is deemed unsustainable.[13]

These efforts have changed the fishing industry forever. Sourcing from fisheries far away that are not under pressure is now the environmentally responsible thing to do. Using the most modern freezing methods and transporting that fish by container ship can even lower the carbon footprint of the entire process. However, this may require substituting threatened local favorites with entirely different fish. There is also an incredibly complicated system of international quotas designed to limit the catches of threatened fish, but the role of wholesalers in overfishing should not be forgotten. It is far easier for fish dealers in any era to question the idea that their stocks are depleting than to plan for a future with fewer or no fish. If wholesalers control the way that fish are sold, it becomes much harder for consumers to understand the environmental implications of what they're eating.

Even though seafood is a global commodity, different countries sell it in different ways. In China, many seafood restaurants keep live fish tanks out where diners can see them. If you order fish, they will take one out of the tank, show it to you before it goes to the kitchen, then serve it with the head still attached. There are cultural explanations for this practice, and flavor-related reasons too (fish heads are, in fact, good eating). However, keeping live fish also serves as quality control. A fish that you can see swimming around before dinner and then identify on your plate is guaranteed fresh. A frozen fish shipped in from halfway around the world may very well be subpar. Shortly before his death, Anthony Bourdain credited the sushi craze for educating Americans about quality fish, saying, "We know now what good fish is, and the market has had to respond to that."[14] Technology may finally be catching up to improvements in customers' tastes, dramatically narrowing the gap between freshly killed fish and the frozen article.

Onboard freezing technology is now good enough that it can be very difficult to distinguish fresh fish from fish that has been "Frozen At Sea" (FAS). FAS means that fish is cleaned on the boat, then frozen at 40 degrees below zero.[15] While customers remain suspicious of any frozen fish, when this is done with care, that suspicion is no longer justified.[16] Effective freezing opens up the possibility of moving high-quality fish anywhere with time no longer an issue, so they can be packed onto frozen shipping containers and sustainably transported across the world. When fish is like any other article in global commerce without any noticeable deterioration, it can become completely commodified. While this may be good news for consumers who want lower prices, it is bad news for the future of any particular fish market because it allows more competition. Like the farm-to-table movement for meat and produce, the combination of the Internet and the indefinite preservation of fish makes it possible for anyone to acquire quality fish from almost anywhere, cutting out multiple links in older provisioning chains.

That competition explains why the New Fulton Fish Market has largely gotten out of the business of directly receiving local fish. During the early days of the original market, the seafood wholesalers created entire provisioning chains with themselves at the center. Now, the wholesalers mostly sell expensive fish from very far away to high-end restaurants and retailers. For example, the global market for shark fins emerged in the 1970s when demand increased in Asia, mainly China and Hong Kong. Sharks had been caught on an incidental basis, but that market combined with the increase in Chinese immigrants in New York seeking fins led to fishermen specifically targeting sharks for the first time. In 1998, the United States exported forty-four tons of shark fin. In 1999, the Fulton Fish Market sold seven different kinds of shark.[17] Every time a particular species was targeted for eating anywhere in the world, it became depleted just a few years later. As a result, shark populations have dropped significantly in all the world's oceans.[18]

The COVID-19 pandemic only added to the wholesalers' woes. Long provisioning chains are susceptible to disasters at every link, and COVID created problems with both supply and demand. Before the pandemic, two-thirds of all the seafood eaten in the United States was consumed in restaurants.[19] Restaurants cut back on orders when most customers stayed home. Obtaining seafood also became harder because fewer boats set sail to catch fish. The Centers for Disease Control recommended that fishermen quarantine for fourteen days before boarding a fishing boat and that various changes in set-up and scheduling be made once on board to enable social distancing. Fish sales in grocery stores actually grew sharply, but that was not enough to make up for the huge losses caused by the loss of restaurants as an end point in the provisioning chain. Sales continued higher well into the pandemic, giving some cause for optimism, but smaller wholesalers seldom deal with supermarkets.[20]

During the first six months of the pandemic, revenue in the entire seafood industry dropped 29 percent. Fish wholesalers, many of whom were already struggling, struggled more because the restaurants that bought their product had trouble paying their debts.[21] "My business has been completely compromised, to a huge degree. I have worked my whole life to build up my business and last year was our best year, so to see it [decline] without anything I can do about it is harmful to my soul," explained a long-active Fulton Fish Market dealer to the website SeafoodSource. High expenses like rent, infrastructure, and insurance remained even as regular commercial purchasers stopped placing orders.[22] The key to economic survival became shortening the provisioning chain so that wholesalers could sell directly to consumers.

The pandemic contributed to the further decentralization of the industry, which had begun long before the old fish market relocated to the Bronx. There is now a company inside the New Fulton Fish Market that will ship fresh fish anywhere in America. In 2019, the CEO bragged, "I can get a fish to Warren Buffett in Omaha, Nebraska, that's as fresh as if he'd walked down to the pier and bought it that morning."[23] Once again, fish from the market could supply the whole country. However, there are plenty of other firms that can do the same. The Internet, like the telephone before it, allows fishermen to talk directly with restaurateurs and other large buyers, cutting out wholesalers entirely. As wholesalers in other cities developed chains of their own that avoided New York City, Fulton Market lost the power to create markets by itself. Bluefin tuna, so prized that it has been fished almost to extinction, might be caught in the Mediterranean, flown to Tokyo for grading and pricing, then flown to Los Angeles for consumption.[24] Moving fish of any kind around the world is relatively easy, so no particular point on the provisioning chain has much more importance than any other as the limits of fish stocks are tested by fishing at the same time in every ocean around the world where popular fish remain.

Since the mid-nineteenth century, more than 90 percent of large spawning fish like cod, haddock, and herring have disappeared.[25] By one assessment, over 80 percent of the world's wild fish stocks have collapsed or cannot safely be fished any longer.[26] Wholesalers have usually avoided the impact of overfishing by sourcing their fish from farther away. In the nineteenth century, that meant Canada or even Alaska. By the late twentieth century, farther away could be any point in the world. However, fish move, migrating seasonally. Getting fish from farther away might mean depleting the same species at different stages in their yearly travels. Some species, like bluefin tuna, cross oceans. Sourcing them from farther away will still likely contribute to their depletion. So what will happen when everywhere in the world becomes depleted simultaneously? We are finding out the answer to that question right now.

The fate of local species in the northeastern United States has also grown more dire. The New York Department of Environmental Conservation banned the commercial and recreational catching of the American shad in 2010 because of severe depletion of the local stock.[27] During the 1990s, the once-dominant cod was replaced on the Georges Bank (off Cape Cod) by dogfish.[28] The eastern oyster has declined by 80 percent since colonization, and the oyster industry, despite a revival in the 1990s, is handling only 14 percent of the business it did at its peak.[29] Wholesale fish dealers saw this problem as it developed and kept right on providing an outlet for the sale of exploited fish. Now, as climate change and warming oceans have contributed to the problem of dwindling fish supplies, restoring stocks has become even more difficult.

Local fishermen have an interest in protecting the health of the fisheries that they catch. Industrial trawlers that can catch fish anywhere in the world don't. Wholesalers can buy from either. Unfortunately, many do not have ties to the regions where their fish originated. The dealers who buy from anywhere and everywhere continue to benefit from the sale of fish, even as stocks of local fish disappear completely. Private benefit has come with a huge public cost. That's why some governments have begun to regulate not just the catching of endangered fish but also their sale. For example, in 2012, New York State banned the possession, sale, and trade of shark fins, essentially declaring the trade in sharks illegal.[30] The dealers in shark were interested in making money, while New York and other states interested in protecting shark species cared about more than the self-interest of a few economic actors, even if the survival of shark species worldwide had little direct effect on them. Areas with local fish can become more sustainable by tapping into those stocks, but despite improvements in water quality in recent decades, local fish will never be able sustain a population as large as New York City again. Consumers of fish becoming more environmentally aware poses a problem for the city's wholesale fish dealers.

Eating fish without guilt requires knowing where the fish on your plate originated and the overall health of the fishery in question. Cutting out the middleman entirely makes it much easier to ensure that your dinner is sustainable. Cutting out the middleman also means cutting out the party that has traditionally stood between the consumer and accurate information about the catch they're eating. Even if you aren't buying directly from the fisherman, there is more information available than ever about which fish choices are environmentally friendly and which are not. Most notably, the Monterey Bay Aquarium has become famous for distributing cards labeling particular species as "avoid," "good alternative," "best choice," and "certified." Seafood Watch, its organization to promote globally sustainable seafood, writes, "From small family-run shrimp farms in Vietnam to large tuna fishing fleets off the Atlantic coast, every seafood product has a story to tell. Knowing the details of how and where your seafood is harvested is key to protecting our oceans and ensuring a long-term supply of seafood."[31] Environmentally minded people take comfort in this information because they have other concerns besides the profit motive.

Seafood Watch is right to suggest that all fish now have a story. Is it farmed or wild caught? Was it caught on a line or in a trawl net? Is that fishery sustainable? Has the fish been accurately identified? Does its name even mean anything? The act of passing fish from one link of the provisioning chain to another increases the chance that its origins will be obscured and the chance the species itself will be obscured, since one fillet of fish can be very difficult to differentiate from another. As was the case at Fulton Market earlier in its history, when someone up the line can make a profit through misidentifying the fish when they sell it, they probably will.[32] To a great extent, the mere existence of seafood wholesalers obscures this kind of information and the stories that go with it. It is very hard for any institution that buys and sells in bulk to know exactly what it has. This explains why the new market for fish involves shipping small amounts directly to people who want to buy it, wherever they happen to be. This development comes as fewer places have many local fish left to catch.

Conservation through regulation has always been an option. In the nineteenth century, markets (including seafood markets) were subject to regulation of all kinds.[33] At the Fulton Fish Market, the most obvious form was sanitary codes that determined how clean stands had to be and how old a fish could be and still be sold. Local regulations concerning where and when fish could be caught became national in scope by the late twentieth century, but they were aimed at fishermen rather than middlemen. Even after the Fulton Fish Market became an entirely wholesale operation, its Lower Manhattan location gave the public the ability to follow how the overall market for fish operated and in some cases, propose regulation. The move to the Bronx represents the near complete abandonment of the public function of a public market. Yes, the City of New York owns the building, but plenty of private institutions are more than capable of providing its citizens with fish of all kinds. In a remote location, with no subway station nearby, the operations do not extend into the adjoining neighborhood, which makes what was once a very public market more like infrastructure than a part of the city's greater cultural landscape.

The original Fulton Fish Market was obviously a market in the sense that it was a place to buy and sell fish, but its long-term historical significance derives more from the other sense of the word "market," namely the abstract idea that there is a set of dedicated buyers for the good that gets sold there. The wholesalers who ran the Fulton Fish Market expanded the scope of the abstract market in order to keep their physical market going. As long as the abstract definition existed, the physical market could be as exploitive and comparatively inefficient as its constituent companies wanted it to be. Nobody really cared about the public good as long as they were all still making money. Because it was the first and most important fish market at the time when industrial fishing was just getting going, the actions of the wholesalers who operated there spurred the general indifference of the industry to the problem of overfishing, despite the obvious cost of this behavior to the amount of fish in the sea.

Thanks to the quality of modern refrigeration, it is theoretically possible that fish from the same species could taste exactly the same, wherever or whenever they were caught. Even if particular species of fresh fish taste the same across time, changes in the state of any fishery are important to the business model of the people selling them. Because more and more fish processing is done by the same firms that are selling it (whether on the boat or not), large conglomerates have an interest in keeping fishing sustainable. Even retailers like Walmart have begun to use their monopoly power as large buyers to change the way popular fisheries are managed by requiring sustainability.[34] Without much attention focused on them, large wholesalers can still take advantage of modern refrigeration and transportation to buy from anywhere, no matter how well that fishery is managed. In some cases, government intervention, like New York's laws against selling shark, have helped limit the effects of this tendency. However, now that all fish are bought and sold in a global marketplace, evading regulation at some point along the provisioning chain has become increasingly easy.

All the way from 1822 down to today, the parties who organized the system for providing fish made those chains longer and more efficient in order to keep them going profitably despite the environmental damage that expanded fishing caused. In the short term, this made sense for economic reasons, but the continued growth of the market for fish of all kinds has only made the inevitable environmental disaster more obvious. Everywhere around the world, big fish that are fished out are replaced by smaller fish that are temporarily abundant because their predators have disappeared. When fishermen can't travel far enough to find abundant stocks of these new species, wholesalers import them from across the world.[35] While seafood wholesalers are not solely to blame for the ongoing disaster that this shortsightedness has caused, when the oceans are so depleted of fish that only the least appealing species remain, we should look to places like the Fulton Fish Market to see where this attitude originated. To begin to rebuild fisheries around the world, it will be necessary to jettison this particular historical legacy in its entirety.

ACKNOWLEDGMENTS

I picked up a copy of Joseph Mitchell's *Up in the Old Hotel* in a bookstore in my hometown of Princeton, New Jersey, shortly after it first came out in 1992. I was immediately transfixed. It wasn't just the quality of the writing, which evoked a mood of a time and place better than anything I had ever read. [If you liked this book and haven't read Mitchell's work yet, you should do so as soon as possible.] It's like a social history time machine, and it convinced me that every single corner of New York City had a history worth uncovering. After I finished my PhD and began researching topics that weren't directly related to the history of the city, I kept reading books on this subject for fun. Even after accumulating a fairly large library on the history of New York, I kept coming back to Mitchell.

Audra Wolfe explained to me that it's best to justify your research through what you've done before, so that I might have something to add to all those existing books about New York. My work on the history of refrigeration became my justification for writing about what might be the most perishable food possible. I didn't originally expect to write all that much about refrigeration here, but since very few people are willing to eat spoiled fish, maybe I should have anticipated things differently. I knew Mitchell loved visiting the Fulton Fish Market but had no idea he wanted to write a history of the place until after I started my research.

Besides Mitchell, the most important chronicler of the Fulton Fish Market is Barbara Mensch. While it was daunting to show my research to someone who witnessed a good chunk of that history with her own eyes, she was extraordinarily supportive of this project, and for that I am incredibly grateful. Besides permission to use a few of her photographs here, she gave me some terrific source advice for chapter 12.

Elizabeth Pillsbury wrote a fantastic dissertation on the history of sport fishing around New York City. She also shared some really useful articles on oysters with me. To find another person interested in typhoid contamination of oysters at the turn of the twentieth century on LinkedIn more than justifies my presence on that platform. When I originally met Martina Caruso of the South Street Seaport Museum, I completely wasted her time for about three hours. She not only graciously talked to me again, she probably gave me more information dealing directly with the fish market than anyone else who helped me with this project. Thank you also to Michelle Kennedy of the same museum for gathering everything I needed to see there. Although the South Street Seaport Museum has never devoted as much attention to the fish market as it has to the port activities that dominated the neighborhood for most of the nineteenth century, I couldn't have written this book without the enormous amount of research and writing that their staff has done on this subject over the last five-plus decades. I got more general historical advice about research and writing from Steven Jaffe, Gergo Baics, and Matthew Morris Booker.

Of course, I am also grateful to my editor, Jennifer Crewe, for her advice, her unflagging support, and teaching me the importance of the serial comma. Columbia University sent my proposal, and eventually the draft manuscript, to three reviewers in total. Their reviews were all incredibly useful. While I may call myself a food historian at this point in my career, fish is a unique subset of that category, and I don't think I could possibly have jumped into the history of New York City like I did here without the help of people who know far more about that subject. After all, reading Joseph Mitchell can only get you so far.

Two research innovations—one new, one old—that made this project possible are databases and interlibrary loan. With respect to the first, many thanks to Thea Lindquist and Pat Talley, who got me onto Hathitrust at CU-Boulder for a very important afternoon. The University of Denver is particularly kind about sharing its digital resources in this area with the public. Bernard Nagengast, who knows far more about the history of refrigeration than I do, pointed me to a bibliography of articles about fish-freezing technology. Then, like with so many other of my research needs, I ordered most of the components of that list with the help of CSU-Pueblo interlibrary loan superstar Kenny McKenzie. ILL is a godsend during ordinary times, but it seemed like a miracle just being able to get articles during the worst days of the pandemic. My particular thanks to Triza Crittle of the National Agricultural Library for scanning that giant history of the New York fisheries article from 1929 when it was hard to get anything else during the COVID-19 lockdown. When librarians returned to campus before most of the rest of us and made books available again through ILL, I wanted to stand up and cheer. I simply couldn't have written this at all without all the librarians who contribute to that whole system. I wish it hadn't taken a pandemic for me to appreciate you as much as you deserve.

In Pueblo, the Provost's Office, John Williamson, Wendy Fairchild, and Lori Blase got me back to New York City as soon as the world started to open up again. A combination of my love of writing and my son Everett kept me sane during lockdown. Like New York City, Pueblo is a place where there is history around every corner. Both places have inspired my interest in urban history, and this book is the first result of my research in this area.

This book is dedicated to my father, who was a native New Yorker.

A NOTE ON SOURCES

This book relies mostly on media, most notably newspaper and magazine articles, about the Fulton Fish Market. When I started it, I had hoped that this might be a good way to examine the lives of the people who worked there across decades, but sadly, most of the reporters who visited were far more interested in fish than people. Although they might mention a few eccentric characters with catchy nicknames like "Carp Carrie" and "Iceberg Tommy," few of these people actually worked at the market when the newspapers made them a little bit famous. Joseph Mitchell befriended people who worked in and around the Fulton Fish Market throughout his long life, but he did not write any profiles of market denizens for *The New Yorker*. Instead, Mitchell wrote about people who worked at different places along the fish provisioning chain, including a fishermen, a fishing boat captain, and a restaurant owner whose establishment specialized in fish.

The exception to that tendency was the "Old Mr. Flood" stories. Justifying his decision to make Hugh G. Flood a composite of many people, Mitchell noted that he had read many other authors' stories about the fish market, but none of those authors ever really understood the "truth" about what happened there.[1] In this book, Joseph Mitchell is both a primary and a secondary source. Mitchell was a people person. Any reporter's story that didn't focus on the people involved with the market would have been unlikely to meet with his approval. But reporters seldom focused their attention upon the people who worked at the market or included the names of those interviewed in a story. Nevertheless, the Fulton Fish Market employed and served hundreds of people whose livelihoods depended upon obtaining a certain understanding of the habits of fish. However, in the total corpus of newspaper articles written about the Fulton Fish Market, there is a lot of truth about the provisioning system with that market at its center, even if the people involved remain something of a mystery.

There are other problems with using these stories as sources. Phillip Lopate's 2007 dismissive assessment of this corpus of fish market reports centers on their repetitiveness rather than their failure to focus on people:

> It is curious how much they resemble one another—how much they are exactly the same article. Usually there is some lyrical prose about dawn, and then the noise, the clamor, the rubber boots, the cats and the seagulls feasting on fish scraps, perhaps a little history going back to 1821, a mention of Al Smith, some statistics about fish consumption or yearly sales, a list of varieties sold, descriptions of unloading, several quotes from grizzled veterans, a regretful sigh about recent changes, and a wish that nothing else would change.[2]

Lopate is definitely right about the tropes, but that doesn't mean every story is identical because the market itself was never the same. More importantly, read enough of these stories and you can begin to get a handle on the aspects of the market that reporters found most interesting because of the thick descriptions they wrote. I've used large chunks of the best of these descriptions in this book because they set the scene at the market so well.

No matter what you think of the literary quality of reporting about the fish market, this observation from Lopate is already outdated: "Dozens of such stories came to be written and now molder in the files of libraries and newspaper morgues."[3] It's not dozens, it's hundreds, and I know that because they are almost all now online and searchable in a variety of very powerful databases. That's how I could spend a one-semester sabbatical at the height of the COVID-19 pandemic writing my first draft of this book without leaving my house. I had taken two research trips to New York City before that, so I had some archival material already (and got a lot more later), but those newspaper and magazine stories are still the best descriptions of the market available and therefore form the heart of this book. They are supplemented by many of the government documents covering topics like fisheries and refrigeration and even the retail fish industry that are now available online in their entirety. Combining this with the miracle that is interlibrary loan, I could do more research than I ever imagined from three-quarters of a continent away.

Supplementing all those stories with archival research, the most important item I found was a 1974 memo in the South Street Seaport Museum archives that explained the history of the various structures that made up the Fulton Fish Market. That memo is the source for the map at the front of this book. Reporters invariably called everything "Fulton Market" or "Fulton Fish Market," including private fish-related structures in the neighborhood, so telling one from another can be very difficult. I had my ideas based on my reading, but reading Ellen Rosebrock's clear description from almost half a century ago was incredibly helpful.

Apart from that invaluable memo, the South Street Seaport Museum has about five boxes of nothing but fish market materials. They includes some primary source material on the fish market in its archive, but more importantly, it has been publishing secondary material about both the market and the fishermen who supplied it for decades now. The Joseph Mitchell Papers at the New York Public Library were obviously very useful, and his article notes for the most fish market-related pieces were most helpful to me. However, the rest of his files resemble a large vertical file from the public library, offering mostly printed material from newspapers and magazines. The New York Municipal Archives has city reports relating to key moments in the history of the fish market and the short (but extremely helpful) description of the place in the WPA's "Feeding the City" manuscript. I've made great use of that work here. A few sources on the market's earliest days came from the collections of the New-York Historical Society.

There are also a few very good article-length histories of the market (other than Mitchell's anecdotal musings) upon which I relied heavily while writing this book. Here they are in alphabetical order:

> Beck, Bruce. *The Official Fulton Fish Market Cookbook*, 3–20. New York: Dutton, 1989.[4]
> Lopate, Phillip. "Introduction." In Barbara Mensch, *South Street*, 1–41. New York: Columbia University Press, 2007.
> Matthews, John H. "A History of the Fisheries Industries of New York." *Fishing Gazette* (Annual Review 1929): 49–80.
> Slanetz, Priscilla Jennings. "A History of the Fulton Fish Market." *The Log of Mystic Seaport* 38 (1986): 14–25.

Together, these sources form a nice summary of the basic parameters of the fish market's operation: when it started, when the buildings went up, how it operated, etc.

This is the first book-length history of the Fulton Fish Market. I think what separates it from these articles is my interest in looking outside the market in order to examine the external forces that shaped its development. Because of the nature of technological and environmental change, if a history of the Fulton Fish Market didn't leave the market more than once in a while, it would be like one of those Greek dramas in which most of the action takes place off stage. Mitchell knew that. I think that's why his relevant *New Yorker* stories mostly took place elsewhere along the provisioning chain.

When I signed on with Columbia University Press to publish this book, my editor, Jennifer Crewe, recommended that I read Theodore Bestor's *Tsukiji: The Fish Market at the Center of the World*, which is about the famous Tokyo fish market, still the only fish market in the world larger than the one in New York City. "Make no mistake," Bestor writes, "the cultural and social details of Tsukiji are central to understanding how the market works."[5] The fact that Mitchell, who was far better steeped in the culture of the Fulton Fish Market than other reporters, felt the need to invent a character who didn't even work there in order to reveal that culture demonstrates that the cultural and social details of the place may well be impregnable.

Following the example of nearly all the reporters who visited this little part of Lower Manhattan over many decades, I wrote primarily about the fish. Of course, the provisioning chain wouldn't have existed without people: people who caught fish, people who transported fish, people who processed fish, and the people who ate it too. I've made a few conjectures based on the available evidence about the culture surrounding fish without necessarily focusing on specific individuals involved in their provisioning. I would have liked to include more about the people who worked at the Fulton Fish Market throughout its history. Nonetheless, I hope there is enough here to illustrate the many changes in all the long fish provisioning chains that shaped the legendary New York institution that was the Fulton Fish Market over the period when it controlled practically every piece of seafood that arrived in the city.

NOTES

INTRODUCTION

1. Calvin Trillin, "Confessions of a Crab Eater," in *The Tummy Trilogy* (New York: Farrar, Straus and Giroux, 1991), 171.
2. Costa Di Mare, accessed September 16, 2019, https://www.wynnlasvegas.com/dining/fine-dining/costa-di-mare.
3. Charles Fishman, *The Wal-Mart Effect* (New York: Penguin Books, 2006), 173–77.
4. On the cattle-to-beef provisioning chain, see William Cronon, *Nature's Metropolis: Chicago and the Great West* (New York: Norton, 1991), 207–59, and Joshua Specht, *Red Meat Republic: A Hoof-to-Tail History of How Beef Changed America* (Princeton, NJ: Princeton University Press, 2019).
5. It was known as "Fulton Market" during most of those years because you could buy a lot more than just fish there. After the retail market for other foods stopped operating, it became generally known as the "Fulton Fish Market." (A Google Ngram of "Fulton Market, Fulton Fish Market" will show the "Fulton Fish Market" line come basically out of nowhere starting around 1920.) Nonetheless, the name "Fulton Market" still appeared often in major New York City publications at the end of the twentieth century. I use the name "Fulton Market" for its early decades, both names for its middle years, and "Fulton Fish Market" for its final decades and when referring to the market in general over its entire history.
6. Evan Osnos, "Old Fulton Fish Market headed for extinction," *Baltimore Sun*, January 14, 2002, https://www.baltimoresun.com/news/bs-xpm-2002-01-14-0201140007-story.html.
7. Priscilla Jennings Slanetz, "A History of the Fulton Fish Market," *The Log of Mystic Seaport* 38 (1986): 20.
8. Molly O'Neill, "A Day Dawns at Fulton," *Newsday*, March 4, 1987, C1.

9. Elizabeth Pillsbury, "An American Bouillabaisse: The Ecology, Politics and Economics of Fishing around New York City, 1870–Present" (Ph.D. diss., Columbia University, 2009), 33.
10. Alfred E. Smith, *Up to Now: An Autobiography* (New York: Viking, 1929), 15–16.
11. Matthew Josephson and Hannah Josephson, *Al Smith: Hero of the Cities* (Boston: Houghton Mifflin, 1969), 46–47.
12. Smith, *Up to Now*, 15.
13. Josephson and Josephson, *Al Smith*, 88.
14. *New York Times*, August 13, 1950, R5.
15. *New York Evening Telegraph*, September 1, 1904, 3.
16. Kathryn Graddy, "Markets: The Fulton Fish Market," *Journal of Economic Perspectives* 20 (Spring 2006): 207–08.
17. Bruce Beck, *The Official Fulton Fish Market Cookbook* (New York: Dutton, 1989), 30.
18. Joseph Mitchell, "Up in the Old Hotel," in *Up in the Old Hotel* (New York: Vintage, 2008), 439. Mitchell retitled a number of his old *New Yorker* stories when he compiled *Up in the Old Hotel* in the early 1990s. "Up in the Old Hotel," for example, was originally called "The Cave." Throughout this book, I cite all of Mitchell's work from there, using those newer titles, because they're much more accessible for anyone who wants to read these brilliant essays than digging them out of old magazines.
19. William Grimes, "A Listener Bends the Ear of Another," *New York Times*, July 22, 1992, C13.
20. There were always a few women working around Fulton Market as buyers, telephone operators, and eventually as sellers, but Mitchell did not write about them.
21. Thomas Kunkel, *Man in Profile: Joseph Mitchell of the New Yorker* (New York: Random House, 2015), 300; Lynn Karpen, "Stand at the Bar and Just Listen," *New York Times Book Review*, August 16, 1992, 7. I make no claim to write as well as Joseph Mitchell might have done had he been able to finish a book on the same subject. However, I have relied on a great deal of Mitchell's research in this work.
22. "I wanted these stories to be truthful rather than factual," Mitchell later wrote of Mr. Flood, "but they are solidly based on facts." Mitchell, *Up in the Old Hotel*, 373.
23. Eleanor Blau, "Fresh Fish Sold Old Way at Fulton Market," *New York Times*, May 28, 1985, B1.
24. Leslie Lindenaur, "The Fulton Fish Market—On Film and on Foot," *Seaport: The Magazine of the South Street Seaport Museum* 18 (Spring–Summer 1984): 6.
25. A. F. Matthews, "New York's Fish Market in Lent," *Harper's Weekly* (March 17, 1894): 255.

1. FISH AND FISHING BEFORE FULTON MARKET

1. Kalm as quoted in Clyde L. MacKenzie, Jr., *The Fisheries of Raritan Bay* (New Brunswick, NJ: Rutgers University Press, 1992), 14.
2. Joseph Mitchell, "The Bottom of the Harbor," in *Up in the Old Hotel* (New York: Vintage, 2008), 469–71. Originally published in 1951.

3. Mark Kurlansky, *The Big Oyster: History on the Half Shell* (New York: Ballantine, 2006), 11–12.
4. Robin Shulman, *Eat the City* (New York: Broadway, 2012), 228–29; Eric W. Sanderson, *Mannahatta: A Natural History of New York City* (New York: Abrams, 2009), 108, 153.
5. MacKenzie, *The Fisheries of Raritan Bay*, 14.
6. William N. Zeisel, Jr., "Shark!!! And Other Sport Fish Once Abundant in New York Harbor," *Seaport* 23 (Winter/Spring 1990): 37.
7. Adriaen Van Der Donck, *A Description of New Netherland*, ed. Charles T. Gehring and William A. Starns (Lincoln: University of Nebraska Press, 2008), 58–59.
8. Sandra L. Oliver, *Saltwater Foodways: New Englanders and Their Food, at Sea and Ashore, in the Nineteenth Century* (Mystic, CT: Mystic Seaport Museum, 1995), 46.
9. Nan A. Rothschild, "Digging for Food in Early New York City," in *Gastropolis: Food and New York City*, ed. Annie Hauck-Lawson and Jonathan Deutsch (New York: Columbia University Press, 2009), 55–56.
10. "New York City's FULTON FISH MARKET," WPA, Federal Writers Project, NYC Unit, 1940, "Feeding the City," Microfilm Roll 131, New York Municipal Archives, New York, NY.
11. Samuel Leavitt, "Fresh Fish," *Appleton's Journal of Literature, Science and Art* 5 (April 1871): 472.
12. Cindy R. Lobel, *Urban Appetites: Food and Culture in Nineteenth-Century New York* (Chicago: University of Chicago Press, 2014), 16.
13. Anders Halverson, *An Entirely Synthetic Fish: How Rainbow Trout Beguiled America and Overran the World* (New Haven, CT: Yale University Press, 2010), 62.
14. James Fenimore Cooper, *The Pioneers* (1823; reprint, New York: Penguin, 1988,), 259, 266.
15. W. Jeffrey Bolster, *The Mortal Sea: Fishing the Atlantic in the Age of Sail* (Cambridge, MA: Belknap, 2012), 31–34.
16. Adam Smith, *The Wealth of Nations* I (Chicago: University of Chicago Press, 1976), 260.
17. Harden F. Taylor, "Refrigeration of Fish," Bureau of Fisheries Document No. 1016 (Washington: Government Printing Office, 1927), 517–18.
18. Trevor Corson, *The Zen of Fish: The Story of Sushi from Samurai to Supermarket* (New York: HarperCollins, 2007), 250.
19. Harold McGee, *On Food and Cooking: The Science and Lure of the Kitchen*, rev. and updated ed. (New York: Scribner, 2004), 203.
20. Sue Shephard, *Pickled, Potted, and Canned: How the Art and Science of Food Preserving Changed the World* (New York: Simon and Schuster, 200), 31, 78, 118–19.
21. Brian Fagan, *Fishing: How the Sea Fed Civilization* (New Haven, CT: Yale University Press, 2017), 166.
22. Charles Sellers, *The Market Revolution: Jacksonian America 1815–1846* (New York: Oxford University Press, 1991), 40.
23. Thomas De Voe was a butcher, a municipal government regulator, and most importantly, a historian of New York City's markets. Any work on the city's markets

during the nineteenth century is dependent upon his books. Although he devotes the most space to matters of meat, he is also a vital source for understanding the changes in the fishing industry throughout the century.

24. Thomas De Voe, *The Market Assistant* (New York: Hurd and Houghton, 1867), 21.
25. Suzanne Wasserman, "Hawkers and Gawkers: Peddling and Markets in New York City," in *Gastropolis: Food and New York City*, ed. Annie Hauck-Lawson and Jonathan Deutsch (New York: Columbia University Press, 2009), 153–54.
26. Thomas F. De Voe, *The Market Book* (New York, 1862), xi.
27. Gergely Baics, *Feeding Gotham: The Political Economy and Geography of Food in New York, 1790–1860* (Princeton, NJ: Princeton University Press, 2016), 279–80, n. 146.
28. Helen Tangires, *Public Markets* (New York: Norton, 2008), 12–13, 25.
29. Helen Tangires, *Public Markets and Civic Culture in Nineteenth-Century America* (Baltimore: Johns Hopkin University Press, 2003), 3, 8.
30. Tangires, *Public Markets*, 25.
31. Tangires, *Public Markets and Civic Culture*, 4, 7; Baics, 20–22.
32. Baics, *Feeding Gotham*, 126.
33. Moreau de St. Mery as quoted in Cindy R. Lobel, "Consuming Classes: Changing Food Consumption Patterns in New York City, 1790–1860" (PhD diss., City University of New York, 2003), 61–62.
34. Richard C. McKay, *South Street: A Maritime History of New York* (New York: G. P. Putnam's Sons, 1934), 252.
35. Egbert Benson, *Memoir Read Before the Historical Society of the State of New York, December 13, 1816*, 2nd ed. (Jamaica: Henry Sleight, 1825), 81.
36. Wasserman, "Hawkers and Gawkers," 154.
37. Baics, *Feeding Gotham*, 74.
38. De Voe, *The Market Book*, 1: 223–24.
39. Baics, *Feeding Gotham*, 128–29.
40. Edwin G. Burrows and Mike Wallace, *Gotham: A History of New York City to 1898* (New York: Oxford University Press, 1999), 354.
41. De Voe, *The Market Book*, 1: 235–36.
42. Benson, *Memoir Read Before the Historical Society of the State of New York*, 81–82.
43. Bruce Beck, *The Official Fulton Market Cookbook* (New York: Dutton, 1989), 5; Preservation and Restoration Bureau, Division of Historic Preservation, "Schermerhorn Row Block: Preliminary Historic Structure Report" (Albany, NY: January 1975), 3; https://babel.hathitrust.org/cgi/pt?id=uiug.30112024570555&view=1up&seq=58.
44. Gina Pollara, "Transforming the Edge: Overview of Selected Plans and Projects," in *The New York Waterfront: Evolution of Building Culture of the Port of New York*, ed. Ken Bone (New York: Monacelli Press, 1997), 158.
45. Ellen Fletcher Rosebrock, *South Street: A Photographic Guide to New York City's Historic Seaport* (New York: Dover, 1977), 43, vii, ix.
46. *New York Times*, December 8, 1881, 10.
47. Ellen Fletcher, *Walking Around in South Street*, rev. ed. (Stony Creek, CT: Leet's Island Books, 1999), 8–9.

48. There were two ordinances to appropriate the necessary land and create Fulton Market. The first came in 1816. The second came in 1817. The only important difference was that the second act included the annexation of the wharves and piers adjoining where the market went. The second act effectively replaced the first because the government decided that it needed to be drawn better. See "The Opinion of Richard Harison and John Wells, Esqs . . .," (New York: Samuel Wood & Sons, 1821), 5–6, from the collection of the New-York Historical Society.
49. Baics, *Feeding Gotham*, 19, 265, n. 3.
50. Fletcher, *Walking Around in South Street*, 45.
51. *Gazette* as quoted in De Voe, *The Market Book*, 1: 488.
52. *Harper's Weekly* 2, January 16, 1858, 41.
53. "To the Honourable the Mayor, Alderman . . .," 1821, New York City Markets Collection, Box 1, New-York Historical Society, New York, NY.
54. *Documents of the Board of Alderman of the City of New York*, vol. 21, part 1 (New York: McSpedon & Baker, 1854), 717; *New York Times*, December 8, 1881, 10.
55. Baics, *Feeding Gotham*, 33, 274, n. 91.
56. *Reports of Cases Argued and Determined in the Supreme Court of Judicature . . . In the State of New York* 20, "In the Matter of the Petition of the Mayor . . . For Enlarging, Extending and Improving Beekman Street . . .," August 1822, 272–73; https://www.google.com/books/edition/Reports_of_Cases_Argued_and_Determined_i/3wU7AQAAMAAJ?hl=en&gbpv=1&dq=For+Enlarging,+Extending+and+Improving+Beekman+Street+1822&pg=PA269&printsec=frontcover.
57. *The Brooklyn Union*, July 2, 1875, 3.
58. *National Advocate*, February 12, 1821.
59. Ellen Fletcher Rosebrock, "Early Market History," *South Street Reporter* 7 (Summer 1973): 7.
60. "Once the new Market was completed in December 1821, stalls were offered at auction. . . . The official opening was January 22, 1822, which accounts for both 'the 1821 Market' and 'the 1822 Market' turning up in historical records." Beck, *The Official Fulton Market Cookbook*, 9. I have seen different dates for the opening of the "Fulton Fish Market" that appear to be timed to the opening of particular buildings, like 1831 and 1882. This is both confusing and incorrect, as fish retailers were in the original from its opening in 1822.

2. THE EARLY DAYS OF FULTON MARKET

1. Edward Ruggles, *A Picture of New York in 1848* (New York: C. S. Francis & Co., 1848), 87–88.
2. John Pintard as quoted in Cindy R. Lobel, *Urban Appetites: Food and Culture in Nineteenth-Century New York* (Chicago: University of Chicago Press, 2014), 18.
3. Bruce Beck, *The Official Fulton Market Cookbook* (New York: E. P. Dutton, 1989), 9.
4. John H. Matthews, "A History of the Fisheries Industries of New York," *Fishing Gazette* (Annual Review 1929): 55.

5. Priscilla Jennings Slanetz, "A History of the Fulton Fish Market," *The Log of Mystic Seaport* 38 (1986): 14.
6. Ellen Rosebrock to Bronson Binger, "Fulton Fish Market: History of Site Occupation and Construction Data," November 14, 1974, in "FULTON FISH MARKET—BUILDINGS," South Street Seaport Museum, New York, NY.
7. Thomas De Voe, *The Market Book* (New York, 1862), 488–89.
8. Gergely Baics, *Feeding Gotham: The Political Economy and Geography of Food in New York, 1790–1860* (Princeton, NJ: Princeton University Press, 2016), 164.
9. Slanetz, "A History of the Fulton Fish Market," 14.
10. *New York Herald*, February 5, 1853, 2.
11. *New York Herald*, February 5, 1853, 2; William Grimes, *Appetite City: A Culinary History of New York* (New York: North Point Press, 2009), 29.
12. Theodore Gill, "On the Fishes of New York," April 14, 1856, in *The Miscellaneous Documents of the Senate of the United States . . . 1856-'57* (Washington, DC: A. O. P Nicholson, 1857), 254. Yes, there was a prominent nineteenth century ichthyologist named Theodore *Gill*. One can only wonder how often someone pointed out to him the appropriateness of his name.
13. *New York Times*, June 14, 1858, 3.
14. *Frank Leslie's Illustrated Weekly*, December 22, 1855, 13.
15. Ellen Fletcher, *Walking Around in South Street*, rev. ed. (Stony Creek, CT: Leete's Island Books, 1999), 8, 12; Kurt C. Schlichting, *Waterfront Manhattan: From Henry Hudson to the High Line* (Baltimore: Johns Hopkins University Press, 2018), 8–9.
16. Alexander Mackay, *The Western World* (London: Richard Bentley, 1849), 78.
17. Ellen Fletcher Rosebrock, *South Street: A Photographic Guide to New York's Historic Seaport District* (New York: Dover, 1977), ix.
18. Joan H. Geismar, "Digging Into the Seaport's Past," *Archeology* 40 (January/February 1987): 35.
19. *New York Herald*, February 5, 1853, 2.
20. *New York Herald*, February 5, 1853, 2.
21. "A Law to Regulate Fulton Market . . .," (New York: G. L. Birch, 1821), 4–12, New-York Historical Society, New York, NY.
22. "A Law to Regulate Fulton Market . . .," 8–9.
23. *Documents of the Board of Alderman of the City of New York*, vol. 21, part 1 (New York: McSpedon & Baker, 1854), 718–19.
24. Baics, *Feeding Gotham*, 1, 36.
25. Baics, *Feeding Gotham*, 199, 36.
26. *New York Tribune*, August 19, 1883, 10.
27. Market Committee, "The ground plan of the thirteen markets of the City of New York," Map Collection, New-York Historical Society, New York, NY.
28. Roger Horowitz, "The Politics of Meat Shopping in Antebellum New York City," in *Meat, Modernity, and the Rise of the Slaughterhouse*, ed. Paula Young Lee (Lebanon: University of New Hampshire Press, 2008), 169, 173.
29. De Voe, *The Market Book*, 511–13.

30. Suzanne Wasserman, "Hawkers and Gawkers: Peddling and Markets in New York City," in *Gastropolis: Food & New York City* (New York: Columbia University Press, 2009): 154–55.
31. De Voe, *The Market Book*, 516.
32. *New York Herald*, February 5, 1853, 2.
33. F. V. Emerson, "A Geographical Interpretation of New York City, Part II," *Bulletin of the American Geographical Society* 40 (1908): 726.
34. Fletcher, *Walking Around*, 4.
35. *The Literary World* 184 (August 10, 1850): 116.
36. Preservation and Restoration Bureau, Division of Historic Preservation, "Schermerhorn Row Block: Preliminary Historic Structure Report" (Albany, NY: January 1975), 13; https://babel.hathitrust.org/cgi/pt?id=uiug.30112024570555&view=1up&seq=58.
37. Edward K. Spann, *New Metropolis: New York City, 1840–1857* (New York: Columbia University Press, 1981), 121–23.
38. Helen Tangires, *Public Markets* (New York: Norton, 2008), 283.
39. Lobel, *Urban Appetites*, 18.
40. Jared N. Day, "Butchers, Tanners and Tallow Chandlers: The Geography of Slaughtering in Early-Nineteenth-Century New York City," in *Meat, Modernity, and the Rise of the Slaughterhouse*, ed. Paula Young Lee, 192–93.
41. *New York Herald*, February 5, 1853, 2.
42. William Grimes, *Appetite City: A Culinary History of New York* (New York: North Point Press, 2009), 29.
43. Edwin G. Burrows and Mike Wallace, *Gotham: A History of New York City to 1898* (New York: Oxford University Press, 1999), 460–61.
44. *New York Tribune*, August 19, 1883, 10.
45. Jane Ziegelman, *97 Orchard: An Edible History of Five Immigrant Families in One New York Tenement* (New York: Harper, 2010), 71.
46. Stephen Buckland, "Eating and Drinking in America: A Stroll Among the Saloons of New York," *Macmillan's Magazine* 16 (October 1867): 462. Emphasis in original.
47. Lobel, *Urban Appetites*, 136–17.
48. *New York Herald*, December 25, 1868, 8.
49. *New York Times*, June 9, 1879, 8.
50. Gill, "On the Fishes of New York," 254.
51. Lobel, *Urban Appetites*, 76.
52. Joseph Mitchell, "Dragger Captain," in *Up in the Old Hotel* (New York: Vintage, 2008), 538; Matthews, "A History of the Fisheries Industries of New York," 56.
53. *Illustrated New York: The Metropolis of To-Day* (New York: International Publishing, 1888), 247; United States Senate Committee on Fisheries, "Fish and Fisheries on the Atlantic Coast," U.S. Senate Report No. 706, 48th Congress, First Session, Serial Set Volume 2179 (Washington: Government Printing Office, 1884), 252.
54. Matthews, "A History of the Fisheries Industries of New York," 56.
55. Carol W. Kimball, "A Country Boy Who Made Good," *The* [New London, CT] *Day*, E3.3, July 1986, 20.

56. *History of Middlesex County, Connecticut* (New York: J. B. Beers & Co., 1884), 366.
57. Maurice E. McLoughlin, "Taught Public to Eat New Kinds of Fish," *Brooklyn Daily Eagle*, July 11, 1932, 15.
58. *New York Sun*, April 7, 1895, 6.
59. *New York Times*, March 3, 1869, 2.
60. *New York Times*, March 8, 1870, 8.
61. "Charter and By-Laws of the Fulton Market Fishmongers' Association of the City of New York" (New York: Slote & Janes, 1869), 10. New-York Historical Society, New York, NY.

3. FISH FROM FAR AWAY

1. Julius W. Muller and Arthur Knowlson, *Deep Sea Fishing Grounds: Fire Island to Barnegat* (New York: Knowlson & Muller, 1915), 21.
2. Thomas De Voe, *The Market Assistant* (New York: Hurd and Houghton, 1867), 203–04.
3. *New York Times*, August 25, 1872, 3.
4. *New York World*, November 23, 1888, 3.
5. Susanne Freidberg, *Fresh: A Perishable History* (Cambridge, MA: Belknap, 2009), 238.
6. Harold McGee, *On Food and Cooking: The Science and Lure of the Kitchen*, rev. and updated ed. (New York: Scribner, 2004), 189.
7. Jonathan Rees, *Refrigeration Nation: A History of Ice, Appliances and Enterprise in America* (Baltimore: Johns Hopkins University Press, 2013), 15–16.
8. Cindy R. Lobel, *Urban Appetites: Food and Culture in Nineteenth-Century New York* (Chicago: University of Chicago Press, 2014), 46.
9. Henry A. Reeves, "The Commerce, Navigation and Fisheries of Suffolk County," in *History of Suffolk County* (Babylon, NY: Budget Steam Print, 1885), 67.
10. *New York Mirror*, September 23, 1826, 71.
11. Raymond McFarland, *A History of the New England Fisheries* (New York: D. Appleton and Co., 1911), 256.
12. Charles H. Stevenson, *The Preservation of Fishery Products for Food* (Washington, DC: Government Printing Office, 1899), 359.
13. *New York Times*, April 6, 1884, 4.
14. Jane Ziegelman, *97 Orchard: An Edible History of the Immigrant Families in One New York Tenement* (New York: Harper, 2010), 71.
15. Marion Harland, *In the Household: A Manual of Practical Cookery* (New York: Scribner, Armstrong & Co., 1873), 51–52.
16. Pierre Blot, *Handbook of Practical Cookery for Ladies and Professional Cooks* (New York: D. Appleton and Company, 1868), 145, 125.
17. *New York Times*, June 4, 1888, 2.
18. De Voe, *The Market Assistant*, 181.

19. John H. Matthews, "A History of the Fishery Industry of New York," *Fishing Gazette* (Annual Review 1929): 73.
20. Sandra L. Oliver, *Saltwater Foodways: New Englanders and Their Food, at Sea and Ashore, in the Nineteenth Century* (Mystic, CT: Mystic Seaport Museum, 1995), 336–37.
21. *Harper's Weekly* 31 (April 30, 1887): 311.
22. *New York Sun*, July 29, 1883, 6. A boat that transported lobster in a fish well was called a lobster smack. A boat that transported fish in a fish well was a fishing smack. A smack may have also been a schooner or a sloop, but a schooner or a sloop was not a smack unless it had a fish well.
23. Joakim Anderssen, "Report on the Department of Fisheries in the World's Exposition in Philadelphia, 1876," in United States Commission of Fish and Fisheries, *Report of the Commissioner for 1878* (Washington, DC: Government Printing Office, 1880), 54.
24. Norman Brouwer, "The Port of New York, 1860–1985: The New York Fisheries," *Seaport* 23 (Winter/Spring 1990): 15–16.
25. *New York Times*, March 9, 1870, 8.
26. William Thomas Okie, "The Thin Ripe Line: Watermelons, Pushcarts, Distribution and Decay," in *Acquired Tastes: Stories About the Origins of Modern Foods*, ed. Benjamin R. Cohen, Michael S. Kideckel, and Anna Zeide (Cambridge: MIT Press, 2021), 84.
27. W. M. P. Dunne, "An Irish Immigrant Success Story," *American Quarterly* 65 (June 1992): 286, 288.
28. United States Commission of Fish and Fisheries, *Report of the Commissioner for 1886* (Washington, DC: Government Printing Office, 1889), 110; *Boston Herald*, September 24, 1886, in *Fishing Scraps* (N.p.: n.p., 1883), 55, Harvard College Library, 1915. This is a one-of-a-kind collection of newspaper clippings about all things fishing. I used the Google Books edition: https://www.google.com/books/edition/Fishing_Scraps/2vt_rEtF86UC?hl=en&gbpv=1.
29. *New York Times*, June 30, 1889, 6.
30. G. Brown Goode and J. W. Collins, "The Fresh Halibut Industry," in George Brown Goode, *The Fisheries and Fishery Industries of the United States*, Section V: History and Methods of the Fisheries, Volume I (Washington, DC: Government Printing Office, 1887), 15.
31. *St. Johns* [Newfoundland] *Evening Telegram*, September 16, 1893, 2. Reprint from the *New York Herald*.
32. G. Brown Goode and J. W. Collins, "The Bank Hand-Line Cod Fishery," in Goode, *The Fisheries and Fishery Industries of the United States*, I: 124–32.
33. Alan Davidson, *North Atlantic Seafood: A Comprehensive Guide with Recipes*, 3rd ed. (Berkeley, CA: Ten Speed Press, 2003), 54, 52.
34. De Voe, *The Market Assistant*, 214.
35. *New York Herald*, February 5, 1853, 2.

36. G. Brown Goode, *American Fishes*, new ed. rev. by Theodore Gill (Boston: L.C. Page and Company, 1903), 337, 348–49.
37. W. Jeffrey Bolster, *The Mortal Sea: Fishing the Atlantic in the Age of Sail* (Cambridge, MA: Belknap, 2012), 137–39, 152–53.
38. *New York Times*, December 31, 1876, 8.
39. *New York Times*, August 11, 1872, 3.
40. Bolster, *The Mortal Sea*, 140.
41. Jeanette [PA] *Daily Dispatch*, February 6, 1893, 2.
42. *New York Tribune*, May 4, 1913, section VI, 4.
43. Henry David Thoreau, *A Week on the Concorde and Merrimack Rivers*, in A Week on the Concord and Merrimack Rivers, Walden; or Life in the Woods, The Maine Woods and Cape Cod (New York: Library of America, 1985), 28–29.
44. Mark Kurlansky, *Salmon: A Fish, the Earth and the History of Their Common Fate* (Ventura, CA: Patagonia, 2020), 117, 128.
45. Oliver, *Saltwater Foodways*, 389.
46. Genio C. Scott, "A Word About Our Fisheries," *Spirit of the Times* 11 (January 8, 1870): 1.
47. De Voe, *The Market Assistant*, 192, 190.
48. *New York Times*, March 31, 1907, 46.
49. Middletown (NY) *Daily Argus*, June 9, 1877, 4. The dealer who did this was Eugene Blackford. He is mentioned below, and there's much more about him in chapter 5.
50. Oliver, *Saltwater Foodways*, 389–90.
51. De Voe, *The Market Assistant*, 200.
52. McFarland, *A History of the New England Fisheries*, 217.
53. *New York World*, May 27, 1888, in *Fishing Scraps*, 24.
54. Louis O. Van Doren, *The Fishes of the East Atlantic Coast* (New York: American Angler, 1884), 91.
55. Alan Davidson, *North Atlantic Seafood*, 3rd ed. (Berkeley, CA: Ten Speed Press, 2003), 35.
56. De Voe, *The Market Assistant*, 200.
57. Mrs. F. L. Gillette, *White House Cookbook* (Chicago: L. P. Miller & Co., 1889), 46.
58. De Voe, *The Market Assistant*, 203.
59. G. G. Huebner, "The Foreign Trade of the United States Since 1789," in *History of Domestic and Foreign Commerce of the United States*, Vol. 2 (Washington, DC: Carnegie Institution, 1915), 194.
60. *Scientific American* 21 (August 7, 1869): 85.
61. Romeo Mansueti and Haven Kolb, *A Historical Review of the Shad Fisheries of North America* (Solomons, MD: Chesapeake Biological Library, 1953), 143.
62. *Good Housekeeping* 2 (February 16, 1886): 216.
63. Sandra L. Oliver, "Saltwater Fish," in *The Oxford Encyclopedia of Food and Drink in America*, Vol. 1, ed. Andrew F. Smith (New York: Oxford University Press, 2004): 483.
64. Louis O. Van Doren, *The Fishes of the East Atlantic Coast* (New York: The American Angler, 1884), 88.
65. Oliver, "Saltwater Fish," 339.

4. THE HEYDAY OF NEW YORK'S OYSTER INDUSTRY

1. *New York Times*, June 9, 1879, 8.
2. Cindy R. Lobel, *Urban Appetites: Food and Culture in Nineteenth-Century New York* (Chicago: University of Chicago Press, 2014), 177–78.
3. Harold McGee, *On Food and Cooking: The Science and Lore of the Kitchen*, rev. and updated ed. (New York: Scribner, 2004), 224, 227–28; Mark Kurlansky, *The Big Oyster: History on the Half Shell* (New York: Ballantine, 2006), 205–06.
4. Clyde L. MacKenzie, Jr., *The Fisheries of Raritan Bay* (New Brunswick, NJ: Rutgers University Press, 1992), 62.
5. John M. Kochiss, *Oystering from New York to Boston* (Middletown, CT: Wesleyan University Press, 1974), 3, 5.
6. Kochiss, *Oystering from New York to Boston*, 24–25, 34; William Grimes, *Appetite City: A Culinary History of New York* (New York: North Point Press, 2009), 44.
7. Joseph Mitchell, "Mr. Flood's Party," in Joseph Mitchell, *Up in the Old Hotel* (New York: Vintage, 2008), 469–71, 418–19. Originally published in 1945.
8. *Paterson [NJ] Daily Press*, October 21, 1872, 1.
9. *The Weekly Herald*, March 12, 1853, 83.
10. David M. Dressel and Donald R. Whitaker, "The U.S. Oyster Industry: An Economic Profile for Policy and Regulatory Analysts," National Fisheries Institute, January 1983, 28.
11. Kurlansky, *The Big Oyster*, 243.
12. *New York Sun*, September 5, 1892, 8.
13. *New York Sun*, September 5, 1892, 8; Clyde L. MacKenzie, Jr., "History of the Fisheries of Raritan Bay, New York and New Jersey," *Marine Fisheries Review* 52, no. 4 (1990): 26.
14. MacKenzie, *Fisheries of Raritan Bay*, 33–34.
15. Andrew F. Smith, "The Food and Drink of New York from 1624 to 1898," in *Gastropolis: Food and Drink in New York City*, ed. Annie Hauck-Lawson and Jonathan Deutsch (New York: Columbia University Press, 2009): 41.
16. Bonnie J. McKay, *Oyster Wars and the Public Trust: Property, Law and Ecology in New Jersey History* (Tucson: University of Arizona Press, 1998), 7–8.
17. MacKenzie, *Fisheries of Raritan Bay*, 24.
18. Ernest Ingersoll, "The Oyster, Scallop, Clam, Mussel and Abalone Industry," in George Brown Goode, *The Fisheries and Fishery Industries of the United States*, Section V: History and Methods of the Fisheries, Volume II (Washington, DC: Government Printing Office, 1887), 521.
19. Joseph Mitchell, "Mr. Hunter's Grave," in *Up in the Old Hotel*, 517. Originally published in 1956. The settlement survived the oyster business, as Mitchell's account of his visit makes clear.
20. Joseph Mitchell, "The Bottom of the Harbor," in *Up in the Old Hotel* (New York: Vintage, 2008), 469–71. Originally published in 1951.
21. *New York Sun*, September 5, 1892, 8.

22. Elizabeth Pillsbury, "Filthy Waters, Fattened Oysters, and Typhoid Fevers: The New York Sewage Battles, 1880–1925," unpublished paper, 5.
23. MacKenzie, "History of the Fisheries," 27.
24. Eugene G. Blackford, "The Oyster Beds of New York," *Proceedings of the American Fisheries Society* 14 (January 1885): 88.
25. Ernest Ingersoll, *A Report on the Oyster-Industry of the United States* (Washington, DC: Government Printing Office, 1881), 88, 6.
26. Kochiss, *Oystering from New York to Boston*, 94; Kurlansky, *The Big Oyster*, 74; Ingersoll, *A Report on the Oyster-Industry*, 249.
27. Kochiss, *Oystering from New York to Boston*, 103, 109.
28. John R. Philpots, *Oysters, and All About Them*, Vol. 2 (London: John Richardson and Company, 1891), 794.
29. *Frank Leslie's Illustrated Newspaper* 49 (December 18, 1879): 263.
30. MacKenzie, *Fisheries of Raritan Bay*, 75.
31. MacKenzie, *Fisheries of Raritan Bay*, 33, 74.
32. Branford [CT] *Opinion*, December 25, 1897, 3.
33. See, for example, Eugene G. Blackford, "Report of the Oyster Investigation and Shell-Fish Commission," New York Assembly Report No. 37 (Troy Press, 1888), 82.
34. *New York Evening World*, August 30, 1888, 2.
35. *New York Times*, September 16, 1883, 2.
36. *New York World*, September 1, 1872, 2.
37. Ingersoll, *A Report on the Oyster-Industry*, 246.
38. Ingersoll, "The Oyster, Scallop, Clam, Mussel and Abalone Industry," 557.
39. *New York Herald*, December 26, 1869, 5.
40. *New York World*, September 1, 1872, 2.
41. *Frank Leslie's Illustrated Newspaper* 43 (February 17, 1877): 391.
42. Fannie Merritt Farmer, *The Boston Cooking-School Cook Book* (Boston: Little, Brown, 1896), 161–64.
43. *New York Herald*, February 5, 1853, 2.
44. Lieut. P. De Broca, "On the Oyster-Industries of the United States," in United States Commission of Fish and Fisheries, *Report of the Commissioner*, 1873-74 and 1874-75 (Washington, DC: Government Printing Office, 1876), 304–05.
45. *New York Herald*, August 22, 1873, 11.
46. Grimes, *Appetite City*, 41.
47. Ernest Ingersoll, *A Week in New York* (New York: Rand McNally, 1891), 286.
48. Mrs. F. L. Gillette, *White House Cookbook* (Chicago: L. P. Miller & Co., 1889), 66.
49. E. P. Churchill, Jr., "The Oyster and the Oyster Industry of the Atlantic and Gulf Coasts," Bureau of Fisheries Document No. 890 (Washington, DC: Government Printing Office, 1920), 45.
50. Sandra L. Oliver, *Saltwater Foodways: New Englanders and Their Food, at Sea and Ashore, in the Nineteenth Century* (Mystic, CT: Mystic Seaport Museum, 1995), 386.
51. *New York Times*, November 8, 1885, 9.

52. *Lloyd's Pocket Companion and Guide Through New York City for 1866–67* (New York: Torrey Brothers, 1866), 120.
53. L. J. Vance, "New York Restaurant Life," *Frank Leslie's Popular Monthly* 35 (January 1893): 106.
54. Grimes, *Appetite City*, 39.
55. Vance, "New York Restaurant Life," 106.
56. *Harper's Weekly* 25 (November 5, 1881): 749.
57. Lobel, *Urban Appetites*, 178.
58. Kurlansky, *The Big Oyster*, 173–77; MacKenzie, *Fisheries of Raritan Bay*, 63.
59. *The New Yorker* 4 (September 22, 1928): 17.
60. MacKenzie, *Fisheries of Raritan Bay*, 63.
61. Kurlansky, *The Big Oyster*, 194.
62. *New York Herald*, December 26, 1869, 5.
63. Lobel, *Urban Appetites*, 131–33.
64. Michael and Ariane Betterberry, *On the Town in New York: From 1776 to the Present* (New York: Charles Scribners' Sons, 1973), 98.
65. Jessica B. Harris, *High on the Hog: A Culinary Journey from Africa to America* (New York: Bloomsbury, 2011), 122–24.
66. Charles Dickens, *American Notes for General Circulation*, Vol. 1 (London: Chapman and Hall, 1842), 208–09.
67. Charles Mackay, *Life and Liberty in America*, Vol. 1 (London: Smith, Elder and Company, 1859), 26.
68. Charles Dickens, *American Notes for General Circulation*, 1: 208–09, 198–99.
69. *The Forest Republican*, November 17, 1880, 1.
70. *Washington Evening Star*, March 20, 1886, 3.
71. Maria Parloa, *Miss Parloa's Kitchen Companion*, 19th ed. (Boston: Estes and Lauriat, 1887), 206.
72. *Omaha Daily Bee*, February 14, 1882, 3.
73. M.F.K. Fisher, *Consider the Oyster* (New York: North Point Press, 1988), 30. Originally published in 1941.

5. THE OPERATION OF A WHOLESALE FISH MARKET

1. John H. Matthews, "A History of the Fisheries Industries of New York," *Fishing Gazette* (Annual Review 1929): 56.
2. *Forest and Stream* 42 (February 17, 1894): 144. The commonality of names demonstrates the importance of family ties in the wholesale fish business. More on this in chapter 12.
3. Benson J. Lossing, *History of New York City*, Vol. 2 (New York: Perine, 1884), 800.
4. Matthews, "A History of the Fisheries Industries of New York," 68–69.
5. Bashford Dean, "Eugene G. Blackford," *Science* 21 (February 10, 1905): 232–33.

6. *Brooklyn Daily Eagle*, May 5, 1879, 2.
7. *New York Sun* (newspaper), *The Sun's Guide to New York* (Jersey City, NJ: Jersey City Printing Company, 1892), 197.
8. Will B. Johnstone, "Fulton Fish Market Centennial Is Here: Old-Timers Recall Its Ancient Glories," *New York Evening Post*, January 7, 1921, 8.
9. Fales Curtis, "Sea Monsters," *Providence Evening Telegram*, February 24, 1889, 12.
10. Dean, "Eugene G. Blackford," 232.
11. *New York Times*, May 29, 1879, 2.
12. *New York Times*, June 9, 1879, 8.
13. Ellen Fletcher Rosebrock, "Early Market History," *South Street Reporter* 7 (Summer 1973): 8.
14. Leslie Shaw, "The Seaport City in 1883," *Seaport: The Magazine of the South Street Seaport Museum* 17 (Summer 1983): 15; *New York Times*, February 10, 1883, 8.
15. George T. Moon, "Fulton Market Greatest Fish Distributing Center of Country," *Fishing Gazette* 35 (July 6, 1918): 976, 978.
16. It is in the collection of the South Street Seaport Museum.
17. *Frank Leslie's Illustrated Newspaper* 11 (December 15, 1860): 10.
18. *Elizabeth City* (NC) *Independent*, February 28, 1936, 1.
19. *The Daily Worker*, December 31, 1928, 4.
20. R. H. Fiedler and J. H. Matthews, "Wholesale Trade in Fresh and Frozen Fishery Products and Related Marketing Considerations in New York City," Appendix VI, *Report of the U.S. Commissioner of Fisheries for 1925* (Washington: Government Printing Office, 1926), 195.
21. Jack Sehres, "A Brief History of the Founding and Organization of the Forwarder in the Fish Industry," in *Less-Than-Carload Freight Traffic and Freight Forwarder Carload Traffic, Hearings Before a Subcommittee of The Committee on Interstate Commerce*, 76th Cong. 3 (Washington, DC: Government Printing Office, 1940), 367.
22. *East Hampton Star*, October 6, 1949, 9.
23. John Gilmer Speed, "Feeding a City Like New York," *Ladies' Home Journal* 13 (July 1896): 4.
24. Alfred E. Smith, *Up to Now: An Autobiography* (New York: Viking, 1929), 15–16.
25. Priscilla Jennings Slanetz, "A History of the Fulton Fish Market," *The Log of Mystic Seaport* 38 (1986): 18; Barbara G. Mensch, *South Street* (New York: Columbia University Press, 2007), 173–74; *Brooklyn Daily Eagle*, November 7, 1886, 2.
26. Dinah Sturgis, "The Great Markets of the World: Fulton Market, New York," *The American Kitchen Magazine* 4 (January 1896): 151.
27. Slanetz, "A History of the Fulton Fish Market," 18.
28. Kathryn Graddy, "The Fulton Fish Market," *The Journal of Economic Perspectives* 20 (Spring 2006): 211–12.
29. *New York Evening World*, July 17, 1919, 13.
30. Russell Maloney, "Fish in Disguise," *New Yorker* 14 (April 2, 1938): 34.
31. *New York Times*, December 20, 1874, 2.
32. *Hartford Courant* (Associated Press), August 12, 1936, 12.

33. *New York Times*, April 23, 1949, 15.
34. Helen Dallas, "Capturing Seafood for Our Tables," *New York Times Magazine*, August 30, 1936, 18.
35. "New York City's FULTON FISH MARKET," WPA, Federal Writers Project, NYC Unit, 1940, "Feeding the City," Microfilm Roll 131, New York Municipal Archives, New York, NY.
36. Sandra L. Oliver, *Saltwater Foodways: New Englanders and Their Food, at Sea and Ashore, in the Nineteenth Century* (Mystic, CT: Mystic Seaport Museum, 1995), 338.
37. U.S. Senate, Committee on Manufactures, *Investigation of Foods Held in Cold Storage* (Washington: Government Printing Office, 1911), 226–27.
38. Fiedler and Matthews, "Wholesale Trade in Fresh and Frozen Fishery Products and Related Marketing Considerations in New York City," 195.
39. City of New York Department of Markets Consumer Services Division, "A Report on Retail Fish Marketing in New York City," December 1938, 2–3. From the Collection of the New York Public Library, New York, NY.
40. Suzanne R. Wasserman, "markets," in *The Encyclopedia of New York City*, ed. Kenneth T. Jackson (New Haven, CT: Yale University Press, 1995), 731.
41. Samuel Leavitt, "Fresh Fish," *Appletons' Journal of Literature, Science and Art* 5 (April 1871): 473.
42. *New York Evening World*, November 23, 1888, 3.
43. Joseph Mitchell, "Interview With Bob Cantalupo," October 26, 1987, 4, Joseph Mitchell Papers, Box 72, Folder 4, no. 1, New York Public Library, New York, NY.
44. *New York Herald*, June 14, 1874, 10.
45. *Monthly Bulletin of the Health Department of the City of New York* 11 (September 1921): 205.
46. *New York Tribune*, January 18, 1901, 10.
47. *The Pittsburgh Press*, July 18, 1902, 17. Reprint from the *New York Evening Post*.
48. City of New York Department of Markets, "A Report on Retail Fish Marketing in New York City," 1.
49. Suzanne Hamlin, "The Secret of Sauce Secreto," *New York Daily News*, September 14, 1983, "Good Living," 9.
50. George T. Moon, "Fulton Market Greatest Fish Distributing Center of Country," *Fishing Gazette* 35 (July 6, 1918): 976.

6. FISHERIES AND THE FISH MARKET

1. H. Bruce Franklin, *The Most Important Fish in the Sea* (Washington, DC: Island Press, 2007), 5–6, 8, 68–70.
2. G. Brown Goode, *A History of the Menhaden* (New York: Orange Judd Company, 1880), 1.
3. *New York Times*, June 23, 1878, 5.
4. *Six Towns Times* [ME], September 14, 1893, 3. Reprinted from the *New York Press*.

5. Goode, *A History of the Menhaden*, 2.
6. Clyde L. McKenzie, Jr., *The Fisheries of Raritan Bay* (New Brunswick, NJ: Rutgers University Press, 1992), 56–57.
7. George Brown Goode, *A Review of the Fishery Industries of the United States and the Work of the U.S. Fish Commission* (London: William Clowes and Sons, 1883), 40.
8. Peter Matthiessen, *Men's Lives* (New York: Vintage, 1986), 145–46.
9. G. G. Huebner, "The Foreign Trade of the United States Since 1789," in *History of Domestic and Foreign Commerce of the United States*, Vol. 2 (Washington: Carnegie Institution, 1915), 176, 200; Franklin, *The Most Important Fish in the Sea*, 93.
10. Huebner, "The Foreign Trade of the United States Since 1789," 200–201.
11. Franklin, *The Most Important Fish in the Sea*, 90–91; *Newark [NJ] Sunday Call*, August 23, 1885, 4.
12. W. Jefferey Bolster, *The Mortal Sea: Fishing the Atlantic in the Age of Sail* (Cambridge, MA: Belknap, 2012), 170. Bolster's book is an absolutely essential source for understanding everything about the depletion of North Atlantic fisheries and its effect upon the environment during this era.
13. *The Fisherman's Own Book* (Gloucester, MA: Proctor Brothers, 1882), 233.
14. Bolster, *The Mortal Sea*, 354.
15. *New York Times*, December 31, 1876, 8.
16. *New York Sun*, July 29, 1883, 6.
17. *The Pittsburgh Press*, August 8, 1907, 15.
18. *Seaport* 26 (Winter 1992): 28.
19. Bolster, *The Mortal Sea*, 118.
20. *Bridgeton [NJ] Pioneer*, July 19, 1883, in *Fishing Scraps* (N.p.: n.p., 1883), 32, Harvard College Library, 1915, https://www.google.com/books/edition/Fishing_Scraps/2vt_rEtF86UC?hl=en&gbpv=1.
21. Bolster, *The Mortal Sea*, 234.
22. *New York Sun*, April 7, 1895, 4.
23. Ted Steinberg, *Gotham Unbound: The Ecological History of Greater New York* (New York: Simon and Schuster, 2014), 172.
24. Joseph Mitchell, "Dragger Captain," in *Up in the Old Hotel* (New York: Vintage, 2008), 540.
25. A. B. Alexander, H. F. Morse, and W. C. Kendall, "Otter-Trawl Fishery," Bureau of Fisheries Document No. 816 (Washington, DC: Government Printing Office, 1915), 13–14.
26. Norman Brouwer, "The Port of New York, 1860–1985: The New York Fisheries," *Seaport* 23 (Winter/Spring 1990): 16.
27. For technical explanations of the differences between trawling methods see Bolster, *The Mortal Sea*, 337, 346, and the description of these methods in his main text, 240–62.
28. Bolster, *The Mortal Sea*, 257–60.
29. *The [New London, CT] Day*, September 25, 1915, 3.
30. John Gilmer Speed, "Feeding a City Like New York," *Ladies' Home Journal* 13 (July 1896): 4.
31. U.S. Senate, *Reports of the Committee on Fisheries, 1884–1887* (Washington, DC: Government Printing Office, 1887), 48, 26–27.

32. *Buffalo Evening Telegraph*, April 30, 1885, 3. Reprinted from the *New York Sun*.
33. Alfred E. Smith, *Up to Now: An Autobiography* (New York: Viking, 1929), 15.
34. *New York Sun*, November 16, 1897, 7.
35. See Steinberg, *Gotham Unbound*, 173; Franklin, *The Most Important Fish in the Sea*, 99; Bolster *The Mortal Sea*, 262.
36. Eugene Blackford, for example, supported some efforts to limit the fishing of species he feared were being seriously depleted but not others. See, for example, *Reports of the Committee on Fisheries, 1884–1887*, 148 and 277.
37. Bolster, *The Mortal Sea*, 221.
38. G. Brown Goode, *American Fishes*, new ed. rev. by Theodore Gill (Boston: L. C. Page and Company, 1903), 462.
39. Anders Halverson, *An Entirely Synthetic Fish: How Rainbow Trout Beguiled America and Overran the World* (New Haven, CT: Yale University Press, 2010), 11, 19, 34.
40. "Blackford, Eugene Gilbert," *The National Cyclopaedia of American Biography* (New York: James T. White and Company, 1891), 394.
41. *New York Sun*, April 2, 1885, 3.
42. Between March 1, 1880, and January 1, 1881, brook trout was the thirty-eighth most popular fish sold at Fulton Market as measured by pounds sold. Salmon trout ranked thirty-third. There were only forty-four different kinds of fish sold at the market that year. See Priscilla Jennings Slanetz, "A History of the Fulton Fish Market," *The Log of Mystic Seaport* 38 (1986): 18; *Brooklyn Daily Eagle*, November 7, 1886, 21.
43. Michael L. Weber, *From Abundance to Scarcity: A History of U.S. Marine Fisheries Policy* (Washington, DC: Island Press, 2002), 3–4.
44. Fred Mather, "Progress in Fish-Culture," *Century Magazine* 5 (April 1884): 900.
45. Elizabeth Pillsbury, "An American Bouillabaisse: The Ecology, Politics and Economics of Fishing around New York City, 1870–Present" (PhD diss., Columbia University, 2009), 21.
46. United States Commission of Fish and Fisheries, *Report of the Commissioner for 1878* (Washington, DC: Government Printing Office, 1880), xlvii, l.
47. Goode, *A Review*, 16–17; Pillsbury, "An American Bouillabaisse," 35. Fish breeding is different than fish farming in that the tiny fish bred in a lab are released in the wild to grow (and oftentimes die). Modern fish farms keep their fish in captivity throughout their lives.
48. *New York Evening Post*, April 28, 1888, 14.
49. Halverson, *An Entirely Synthetic Fish*, 85.
50. John Burroughs, "Our River," *Scribner's Monthly* 20 (August 1880): 492.
51. A. N. Cheney, "Shad Culture in the Hudson River," in *Fifth Annual Report of the Commissioners of Fisheries, Game and Forests of the State of New York* (Albany, 1900), 244–45.
52. *Brooklyn Daily Eagle*, April 9, 1899, 22.
53. Huebner, "The Foreign Trade of the United States Since 1789," 195.
54. Romeo Mansueti and Haven Kolb, *A Historical Review of the Shad Fisheries of North America* (Solomons, MD: Chesapeake Biological Library, 1953), 153.
55. Matthiessen, *Men's Lives*, 84, 133.

56. Pillsbury, "An American Bouillabaisse," 84.
57. Clyde L. MacKenzie, Jr., "History of the Fisheries of Raritan Bay, New York and New Jersey," *Marine Fisheries Review* 52, no. 4 (1990): 18.
58. *Forest and Stream* 42 (February 7, 1894)): 144.
59. George W. Perkins, "How a City Market Department Could Decrease Cost of Food," *Real Estate Record and Builders' Guide* 99 (June 9, 1917): 803.
60. *Commercial Fisheries Review* 26 (November 1964): 40.
61. *Lawrence [KS] Journal-World* (Associated Press), June 23, 1925, 6.
62. "New York City's FULTON FISH MARKET," WPA, Federal Writers Project, NYC Unit, 1940, "Feeding the City," Microfilm Roll 131, New York Municipal Archives, New York, NY.
63. Sarah Tyson Rorer, *Mrs. Rorer's New Cook Book* (Philadelphia: Arnold and Company, 1902), 96.

7. TURTLE AND TERRAPIN

1. *Brooklyn Daily Eagle*, December 22, 1889, 10.
2. Joseph Mitchell, "The Same as Monkey Glands," in *Up In the Old Hotel* (New York: Vintage, 2008), 317–18. The original title for this Mitchell story was "Mr. Barbee's Terrapins."
3. *New York Times*, August 30, 1908, 34. He is not named in the article, but was almost certainly Walter T. Smith. Blackford died in 1904.
4. James J. Parsons, *The Green Turtle and Man* (Gainesville: University of Florida Press, 1962), 26.
5. *New York Evening Post*, July 22, 1830, 3
6. *New York Evening Post*, July 20, 1831, 3.
7. *New York World*, December 17, 1860, 6.
8. *New York Times Magazine*, October 6, 1907, 11.
9. Mrs. H. W. Beecher, "Turtles," *The Christian Union* 22 (October 16, 1880): 308.
10. *New York World*, December 17, 1860, 6.
11. Frederick W. True, "The Turtle and Terrapin Fisheries," in George Brown Goode, *The Fisheries and Fishery Industries of the United States*, Section V: History and Methods of the Fisheries, Vol. II (Washington, DC: Government Printing Office, 1887), 497.
12. *Frank Leslie's Illustrated Newspaper*, January 20, 1883, 359.
13. *St. Louis Republic*, November 21, 1891, 16.
14. Donald K. Tressler, *Marine Products of Commerce* (New York: The Chemical Catalog Company, 1923), 597–98.
15. Paul Freedman, *Why Food Matters* (New Haven, CT: Yale University Press, 2021), 132.
16. *Mrs. Rorer's Philadelphia Cook Book* as quoted in Sandra L. Oliver, *Saltwater Foodways: New Englanders and Their Food, at Sea and Ashore, in the Nineteenth Century* (Mystic, CT: Mystic Seaport Museum, 1995), 184.

17. *The Branford* [CT] *Opinion*, January 19, 1906, 15. Reprint from the *New York Press*.
18. Amy B. Trubek, "Turtle Soup," *Gastronomica* 1 (Winter 2001): 12.
19. *New York Times*, December 18, 1881, 13.
20. "The Hoboken Turtle Club," *My Inwood*, September 21, 2013, https://myinwood.net/the-hoboken-turtle-club/#:~:text=According%20to%20legend%2C%20as%20the,in%20George%20Washington's%20Continental%20army.
21. *Frank Leslie's Illustrated Newspaper*, July 17, 1886, 343.
22. Alison Rieser, *The Case of the Green Turtle: An Uncensored History of a Conservation Icon* (Baltimore: Johns Hopkins University Press, 2012), 15–16, 24.
23. Karen Pinchin, "How Slaves Shaped American Cooking," *National Geographic*, March 1, 2014, https://www.nationalgeographic.com/news/2014/3/140301-african-american-food-history-slavery-south-cuisine-chefs/.
24. *New Orleans Daily Picayune*, September 1, 1894, 6.
25. True, "The Turtle and Terrapin Fisheries," 500.
26. *Washington Star*, June 18, 1911, 7.
27. *New York Evening World*, November 15, 1887, 3.
28. *New York Times*, January 12, 1896, 19.
29. William Temple Hornaday, *The American Natural History* (New York: Charles Scribner's Sons, 1922), 327.
30. *New York Times*, July 18, 1878, 8.
31. True, "The Turtle and Terrapin Fisheries," 500.
32. Mitchell, "The Same as Monkey Glands," 322–23.
33. *New York Tribune*, September 11, 1874, 7.
34. *New York World*, August 18, 1895, 17.
35. *New York Tribune*, March 22, 1896, 32.
36. Christina Tkacik, "Terrapin, the Forgotten Maryland Delicacy," *Baltimore Sun*, June 16, 2016, http://darkroom.baltimoresun.com/2016/06/from-the-vault-terrapin-the-forgotten-maryland-delicacy-3/#1.
37. Robert E. Coker, "The Diamond-Back Terrapin: Past, Present and Future," *The Scientific Monthly* 11 (August 1920): 185.
38. Mitchell, "The Same as Monkey Glands," 314–24.
39. Ernest Freeberg, *A Traitor to His Species: Henry Bergh and the Birth of the Animal Rights Movement* (New York: Basic Books, 2020), 7–10.
40. Freeberg, *A Traitor to His Species*, 7.
41. *New York Herald*, May 23, 1872, 8.
42. *New York Sun*, July 2, 1876, 5.
43. Freeberg, *A Traitor to His Species*, 10–11, 269–70.
44. *New York Times*, May 17, 1885, 6.
45. *St. Tammany Farmer*, June 20, 1903, 2.
46. *New York Times*, August 27, 1895, 8.
47. George Tucker, "Man About Manhattan," *Wilmington* (NC) *Morning Star*, June 7, 1940, 4.
48. Mitchell, "The Same as Monkey Glands," 323–24.

49. Jane Holt, "News of Food; Terrapin Stew Coming Back Into Popularity but in Less Abundance Than in Halcyon Days," *New York Times*, February 8, 1945, 16.
50. *LIFE Magazine* 22 (March 10, 1947): 121.
51. May-bo Ching, "The Flow of Turtle Soup from the Caribbean via Europe to Canton, and Its Modern American Fate," *Gastronomica* 16 (Spring 2016): 87.
52. In New Orleans, both turtle soup and terrapin stew remain staples of local cuisine. Turtle soup is also a staple of Philadelphia cuisine and is still available at places like the Union League, Old Original Bookbinder's, and the Reading Terminal Market. Bookbinder's still sells it in cans. At the Commander's Palace in New Orleans, the legendary turtle soup now depends upon farm-raised alligator snapping turtles. Of course, mock turtle soup remains an option, even if it isn't quite as popular as it once was. See Teagan Schweizer, "The Turtles of Philadelphia's Culinary Past," *Expedition Magazine* 51 (2009), https://www.penn.museum/sites/expedition/the-turtles-of-philadelphias-culinary-past/; and Shoshi Parks, "Our Taste For Turtle Soup Nearly Wiped Out Terrapins. Then Prohibition Saved Them," National Public Radio, July 18, 2019, https://www.npr.org/sections/thesalt/2019/07/18/742326830/our-taste-for-turtle-soup-nearly-wiped-out-terrapins-then-prohibition-saved-them.
53. *LIFE Magazine* 22 (March 10, 1947): 121.
54. True, "The Turtle and Terrapin Fisheries," 498, 500.
55. Freeberg, *A Traitor to His Species*, 271.

8. FREEZING, COLD STORAGE, AND IMPROVEMENTS IN TRANSPORTATION

1. *New York Times*, November 13, 1941, 27.
2. Dale Carnegie, untitled column, *Warsaw [IN] Daily Union*, August 19, 1941, 3. I have altered the punctuation and the spacing in the quotes for the sake of clarity.
3. Bruce Beck, *The Official Fulton Fish Market Cookbook* (New York: Dutton, 1989), 17.
4. Bruce Stutz, "Fish Market Taxonomy," *Natural History* 95 (August 1986): 77.
5. Jonathan Rees, *Refrigeration Nation: A History of Ice, Appliances and Enterprise in America* (Baltimore: Johns Hopkins University Press, 2013), 69.
6. *New York Evening World*, June 30, 1894.
7. *New York Evening World*, August 22, 1888, 2.
8. Charles H. Stevenson, *The Preservation of Fishery Products for Food* (Washington, DC: Government Printing Office, 1899), 371–73.
9. Sandra L. Oliver, *Saltwater Foodways: New Englanders and Their Food, at Sea and Ashore, in the Nineteenth Century* (Mystic, CT: Mystic Seaport Museum, 1995), 341.
10. Stevenson, *The Preservation of Fishery Products for Food*, 373.
11. Rees, *Refrigeration Nation*, 3.
12. Mark Kurlansky, *Birdseye: The Adventures of a Curious Man* (New York: Doubleday, 2012), 118–21.

13. *New York Sun*, March 24, 1907, 1.
14. C. L. Alsberg, "The Relation of Bureau of Chemistry to Sea Foods," *Fishing Gazette* 35 (July 6, 1918): 942.
15. *Industrial Refrigeration* 12 (January 1897): 15. Reprinted from the *Fishing Gazette*.
16. James M. Lemon, "Developments in Refrigeration of Fish in the United States," Bureau of Fisheries Investigational Report No. 16 (Washington, DC: Government Printing Office, 1932), 3.
17. John A. Dassow, S. R. Pottinger, and John Holston, "Refrigeration of Fish: Part Four: Preparation, Freezing and Cold Storage of Fish, Shellfish and Precooked Fishery Products," United States Department of the Interior, Fishery Leaflet 430, 1956, 4.
18. Birdseye's contribution to this process was to introduce a method for freezing fish fillets (processed fish in molds) that didn't require direct contact with the source of the cold. With the help of the federal government's Bureau of Fisheries, this technique dramatically expanded the market for frozen cod fillets. See Michael L. Weber, *From Abundance to Scarcity: A History of U.S. Marine Fisheries Policy* (Washington, DC: Island Press, 2002), 25.
19. Harden F. Taylor, "Refrigeration of Fish," Bureau of Fisheries Document No. 1016 (Washington, DC: Government Printing Office, 1927), 609–11.
20. Trevor Corson, *The Zen of Fish: The Story of Sushi from Samurai to Supermarket* (New York: HarperCollins, 2007), 248.
21. Lemon, "Developments in Refrigeration of Fish in the United States," 14–17.
22. Joakim Andersson, "Report on the Department of Fisheries in the World's Exposition in Philadelphia 1876," in United States Commission of Fish and Fisheries, *Report of the Commissioner for 1878* (Washington: Government Printing Office, 1880): 54.
23. *Meriden [CT] Daily Republican*, 21 February 1887, 4.
24. John H. Matthews, "A History of the Fisheries Industries of New York," *Fishing Gazette* (Annual Review, 1929): 77.
25. *Report of the New York State Food Commission for Period October 18, 1917, to July 1, 1918* (Albany: J.B. Lyon Company, 1919), 37–38; *New York Times*, 1 June 1912, 22.
26. Rees, *Refrigeration Nation*, 103–09.
27. New York, New Jersey Port and Harbor Redevelopment Commission, *Joint Report With Comprehensive Plan and Recommendations* (Albany: J. B. Lyon Company, 1920), 338.
28. Federal Trade Commission, *Report of the Federal Trade Commission on the Wholesale Marketing of Food*, June 30, 1919 (Washington, DC: Government Printing Office, 1920), 208.
29. Rees, *Refrigeration Nation*, 111–12.
30. H. W. Loweree, "A Day at Fulton Market," *Power Boating* 25 (July 1923), 12.
31. Elizabeth Pillsbury, "An American Bouillabaisse: The Ecology, Politics and Economics of Fishing Around New York City, 1870–Present" (PhD diss., Columbia University, 2009), 164.
32. Taylor, "Refrigeration of Fish," 503–04.
33. G. T. Ferris, "How a Great City Is Fed," *Harper's Weekly* 34 (March 22, 1890): 230.
34. *New York Times*, March 31, 1907, 46.

35. Dinah Sturgis, "The Great Markets of the World. Fulton Market, New York," *The American Kitchen Magazine* 4 (January 1896): 152–53.
36. *New York Times*, March 31, 1907, 46.
37. *The [New London, CT] Day*, June 3, 1904, 3.
38. Stevenson *The Preservation of Fishery Products for Food*, 341–42.
39. *New York Tribune*, September 25, 1909, D17.
40. Ann Buttenwieser, "'Fore and Aft': The Waterfront and Downtown's Future," in *The Lower Manhattan Plan: The 1966 Vision for Downtown New York*, ed. Carol Willis (New York: Princeton Architectural Press, 2002), 21.
41. *New York Times Magazine*, April 21, 1912, 7.
42. R. H. Fiedler and J. H. Matthews, "Wholesale Trade in Fresh and Frozen Fishery Products and Related Marketing Considerations in New York City," Appendix VI, *Report of the U.S. Commissioner of Fisheries for 1925* (Washington, DC: Government Printing Office, 1926), 186–87, 192.
43. Madigan-Hyland Consulting Engineers, "Fulton Fish Market Study," Department of Public Works, NYC, 1953, iii, New York Municipal Archives, New York, NY.
44. Raymond McFarland, *A History of the New England Fisheries* (New York: D. Appleton and Co., 1911), 289–90.
45. *Fishing Gazette* 35 (July 6, 1918): 995.
46. *East Hampton Star*, September 16, 1932, 7.
47. James L. Anderson, *The International Seafood Trade* (Boca Raton, FL: CRC Press, 2003), 10.
48. G. C. Eddie, "Recent Developments in the Freezing of Fish at Sea," *Chemistry and Industry* 40 (October 1, 1955): 1249.
49. Dassow, Pottinger, and Holston, "Refrigeration of Fish," 24, 31.
50. Eddie, "Recent Developments in the Freezing of Fish at Sea," 1248–49.
51. Mansel G. Blackford, *Making Seafood Sustainable* (Philadelphia: University of Pennsylvania Press, 2012), 30.
52. William W. Warner, *Distant Water: The Fate of North Atlantic Fishermen* (New York: Penguin, 1983), 37–39.
53. Warner, *Distant Water*, 39, 48–49, 60.
54. Pillsbury, "An American Bouillabaisse," 235.
55. Blackford, *Making Seafood Sustainable*, 14, 31.
56. Carl Safina, *Song for the Blue Ocean* (New York: Holt, 1997), 71.
57. Blackford, *Making Seafood Sustainable*, 156.

9. FROM THE BROOKLYN BRIDGE TO THE FDR DRIVE

1. *Brooklyn Daily Eagle*, January 11, 1912, 22.
2. Ellen Fletcher Rosebrock, *South Street: A Photographic Guide to New York's Historic Seaport* (New York: Dover, 1977), 3

3. Daniel Czitrom, "The Wickedest Ward in New York," *Seaport* (March/April/May 2008): 8.
4. Ellen Fletcher, *Walking Around in South Street*, rev. ed. (Stony Creek, CT: Leete's Island Books, 1999), 32.
5. James M. Lindgren, *Preserving South Street Seaport: The Dream and Reality of a New York Urban Renewal District* (New York: New York University Press, 2014), 105.
6. Madigan-Hyland Consulting Engineers, "Fulton Fish Market Study," Department of Public Works, NYC, 1953, i, New York Municipal Archives, New York, NY.
7. William E. Geist, "About New York: An Old Salt Wonders Why People Pay $20 a Fish," *New York Times*, September 20, 1986, 30.
8. Joseph Mitchell, "Up in the Old Hotel," in Joseph Mitchell, *Up in the Old Hotel* (New York: Vintage, 2008), 458. Originally published in 1952.
9. Miller Hageman, "Stories of Old Fulton Market," *Brooklyn Citizen*, May 21, 1899, 8.
10. *Brooklyn Times Union*, January 20, 1924, 6.
11. Thomas Bender, *The Unfinished City: New York and the Metropolitan Idea* (New York: New York University Press, 2007), 12–13.
12. Harry Hall, "New York Has Not Stopped Growing Which Means That Opportunities Are as Great as Those of Yesterday," *New York Tribune*, December 28, 1919, viii.
13. City of Boston Committee on Markets, "Report of Committee on Markets on the Markets of New York, Philadelphia, Baltimore and Washington," City Document No. 88, May 12, 1902, 18–19.
14. *The Outlook* (October 7, 1911): 339–40.
15. "Report of Committee on Markets," 19.
16. *Forest and Stream* 42 (February 17, 1894): 144.
17. John H. Matthews, "A History of the Fisheries Industries of New York," *Fishing Gazette* (Annual Review 1929): 70.
18. Fletcher, *Walking Around in South Street*, 21.
19. *Fishing Gazette* 135 (July 6, 1918): 983.
20. Martina Caruso, "Seaport Architectural Gems," South Street Seaport Museum, October 1, 2020, https://southstreetseaportmuseum.org/seaport-architectural-gems/.
21. Kurt C. Schlichting, *Waterfront Manhattan: From Henry Hudson to the High Line* (Baltimore: Johns Hopkins University Press, 2018), 121–22.
22. Preservation and Restoration Bureau, Division of Historic Preservation, "Schermerhorn Row Block: Preliminary Historic Structure Report" (Albany, NY: January 1975), 17; https://babel.hathitrust.org/cgi/pt?id=uiug.30112024570555&view=1up&seq=58.
23. *New York Times*, January 13, 1914, 7.
24. *Real Estate Record and Builders' Guide* 99 (May 26, 1917): 732.
25. *Report of the Federal Trade Commission on the Wholesale Marketing of Food*, June 30, 1919 (Washington, DC: Government Printing Office, 1920), 208.
26. Ellen Fletcher, "Saved From the Wrecker's Ball," *Seaport: The Magazine of the South Street Seaport Museum* 17 (Summer 1983): 38.
27. Ellen Rosebrock, "Fulton Fish Market's 'Tin Building,'" *South Street Reporter* (Fall 1974): 19.

28. *New York Tribune*, January 18, 1903, 10.
29. *New York Times*, August 12, 1917, 7.
30. Priscilla Jennings Slanetz, "A History of the Fulton Fish Market," *The Log of Mystic Seaport* 38 (1986): 16.
31. Matthews, "A History of the Fisheries Industries of New York," 59–60.
32. Matthews, "A History of the Fisheries Industries of New York," 61–63.
33. Bruce Beck, *The Official Fulton Fish Market Cookbook* (New York: Dutton, 1989), 15.
34. Edwards and Kelcey, "Wholesale Fish Market for New York City: Survey, Study, Report," New York City Department of Public Works, 1963, 21. New York Municipal Archives, New York, NY. By the late 1980s, there were only two firms left that dealt in freshwater fish, and neither did business on Peck Slip.
35. Charlotte Hughes, "A New Day for the Old Fulton Fish Market," *New York Times Magazine*, April 24, 1938, 12.
36. Walker Evans, "On the Waterfront," *Fortune* 62 (November 1960): 145.
37. Leba Presner, "South Street Perks Up," *New York Times*, June 11, 1939, 16 XX.
38. Hughes, "A New Day for the Old Fulton Fish Market," 12.
39. *East Hampton* (NY) *Star*, September 16, 1932, 7.
40. Emanuel Perlmutter, "Lanza: Case History of a Thug," *New York Times*, April 21, 1957, 7E. There are countless explanations for Lanza's nickname, ranging from his wearing funny socks to the fact that he would "sock" people who didn't pay him. It is impossible to tell which one is the true origin story.
41. Walter Chambers, *Labor Unions and the Public* (New York: Coward-McCann, c. 1936), 144–45.
42. *TIME Magazine*, April 20, 1931, http://content.time.com/time/subscriber/article/0,33009,741409,00.html.
43. Joseph Mitchell, Fulton Fish Market Notes, June 1978, Joseph Mitchell Papers, Box 97, Folder 5; Archives and Special Collections New York Public Library, New York, NY.
44. *The* [New London, CT] *Day*, December 14, 1931, 8.
45. *New York Times*, August 3, 1940, 17.
46. *Brooklyn Daily Eagle*, September 11, 1931, 1.
47. *New York Times*, June 6, 1933, 21
48. *New York Times*, April 25, 1936, 2; Emanuel Perlmutter, "Lanza: Case History of a Thug," *New York Times*, April 21, 1957, 7E.
49. Robert J. Kelly, *The Upper World and the Underworld: Case Studies of Racketeering and Business Infiltrations in the United States* (New York: Kluwer Academic Publishing, 1999), 106.
50. *East Hampton Star*, September 16, 1932, 7.
51. State of New York Conservation Department, "A Biological Survey of the Salt Waters of Long Island, 1938" (Albany: J. B. Lyon, 1939), 44.
52. W. O. Saunders, "Fulton Market Fish Dealers Get an Awkward Break in the News Says This Writer," *Elizabeth City* [NC] *Independent*, December 20, 1935, 5.

53. Saunders "Fulton Market Fish Dealers," 1.
54. *New York Times*, October 11, 1968, 47.
55. Selwyn Raab, *Five Families: The Rise, Decline and Resurgence of America's Most Powerful Mafia Empires* (New York: Thomas Dunne, 2005), 566.
56. "Fulton Fish Market, No. 1 Fish Center, Again Survives a Major Mishap," unidentified newspaper clipping, 1936, Fulton Fish Market Clipping File, Milstein Division, New York Public Library, New York, NY.
57. *Kingston [NY] Daily Freeman* (Associated Press), August 11, 1936, 11.
58. Beck, *The Official Fulton Fish Market Cookbook*, 14; Ellen Rosebrock to Bronson Binger, "Fulton Fish Market: History of Site Occupation and Construction Data," November 14, 1974, in "FULTON FISH MARKET—BUILDINGS," South Street Seaport Museum, New York, NY.
59. Presner, "South Street Perks Up," 16 XX.
60. Beck, *The Official Fulton Fish Market Cookbook*, 15.
61. John Freeman Gill, "A Slice of the Fulton Fish Market Gets a New Life," *New York Times*, February 28, 2020, https://www.nytimes.com/2020/02/28/realestate/a-slice-of-the-fulton-fish-market-gets-a-new-life.html.
62. *New York Times*, April 16, 1969, 95.
63. Phillip Lopate, "Introduction," in Barbara G. Mensch, *South Street* (New York: Columbia University Press, 2007), 14–15.
64. Robert A.M. Stern, Thomas Mellins, and David Fishman, *New York 1960: Architecture and Urbanism Between the Second World War and the Bicentennial* (New York: The Monacelli Press, 1995), 207.
65. Lindgren, *Preserving South Street Seaport*, 13.
66. Robert Sylvester, "Dream Street," *New York Daily News*, April 14, 1960, c16.
67. Joseph Mitchell, "Up in the Old Hotel," 458.
68. Preservation and Restoration Bureau, "Schermerhorn Row Block," 17.
69. "The Talk of the Town: David Rockefeller," *The New Yorker* 36 (July 23, 1960): 16.

10. POLLUTION AND THE DECLINE OF NEW YORK'S OYSTER INDUSTRY

1. *New York Times*, August 28, 1885, 8.
2. Mike Wallace, *Greater Gotham: A History of New York City from 1898 to 1919* (New York: Oxford University Press, 2017), 216–17.
3. William N. Zeisel, Jr., "Shark!!! And Other Sport Fish Once Abundant in New York Harbor," *Seaport* 23 (Winter/Spring 1990): 39.
4. Ted Steinberg, *Gotham Unbound: The Ecological History of Greater New York* (New York: Simon and Schuster, 2014), 77.
5. David Soll, *Empire of Water: An Environmental and Political History of the New York City Water Supply* (Ithaca, NY: Cornell University Press, 2013), 19.

6. David Schuyler, *Embattled River: The Hudson and Modern American Environmentalism* (Ithaca, NY: Cornell University Press, 2018), 81–82.
7. Harrison P. Eddy, "Report of Harrison P. Eddy," in Metropolitan Sewerage Commission of New York, "Present Sanitary Condition of New York Harbor . . .," August 1, 1912 (New York: Wynkoop Hallenbeck Crawford, 1912), 139.
8. Bonnie J. McKay, *Oyster Wars and the Public Trust: Property, Law, and Ecology in New Jersey History* (Tucson: University of Arizona Press, 1998), 155.
9. Elizabeth Pillsbury, "Filthy Waters, Fattened Oysters, and Typhoid Fevers: The New York Sewage Battles, 1880–1925," unpublished paper, 1.
10. *New York Times*, November 20, 1894, 4.
11. Charles P. Chapin, "Report of Charles P. Chapin," in Metropolitan Sewerage Commission, "Present Sanitary Conditions," 121.
12. Pillsbury, "Filthy Waters," 1, n. 1.
13. *Brooklyn Times Union*, December 29, 1902, 6.
14. Pillsbury, "Filthy Waters," 2.
15. *New York Tribune*, February 23, 1908, 15.
16. *Department of Health of the City of New York Annual Report for Year ending December 31, 1908* (New York: Martin. B. Brown Company, 1909), 46.
17. George W. Stiles, Jr., "Sewage-Polluted Oysters as a Cause of Typhoid and Other Gastrointestinal Disturbances," Bureau of Chemistry Bulletin No. 156 (Washington, DC: Government Printing Office, 1912): 14.
18. Stiles, Jr., "Sewage-Polluted Oysters," 29.
19. Stiles, Jr., "Sewage-Polluted Oysters," 28–33.
20. John Waldman, *Heartbeats in the Muck: The History, Sea Life, and Environment of New York Harbor*, rev. ed. (New York: Empire State Editions, 2013), 106.
21. *New York Sun*, April 6, 1916, 11.
22. *The Brooklyn Citizen*, September 13, 1912, 3.
23. *Annual Report of the Board of Health of the Department of Health of the City of New York*, Year Ending December 31, 1912 (New York: 1913), 37.
24. *Annual Report of the Department of Health of the City of New York*, 1919 (New York: 1920), 203.
25. *New York Tribune*, November 4, 1904, 5.
26. John H. Matthews, "A History of the Fishery Industries of New York," *Fishing Gazette* (Annual Review 1929): 67.
27. *New York Sun*, February 3, 1907, 39.
28. *Ithaca Chronicle and Democrat*, Hudson-Fulton Number, September 23, 1909, 17.
29. L. L. Lumsden, H. E. Hasseltime, J. P. Leake, and M. V. Veldee, "A Typhoid Fever Epidemic Caused by Oyster-Borne Infection (1924–1925)," Supplement No. 50 to the Public Health Reports (Washington, DC: Government Printing Office, 1925), 1.
30. *New York Times*, January 2, 1925, 7.
31. Lumsden et al., "A Typhoid Fever Epidemic Caused by Oyster-Borne Infection (1924–1925)," 52.

32. See "Ockers, Jacob," in *American Biography: A New Cyclopedia*, vol. 5 (New York: The American Historical Society, 1919), 205–09. The company continued after Ockers sold it in 1912, went through several name changes, and remained in business until 1999. It was called the Bluepoints Oyster Company at the time of this outbreak. See "Historic Coasts," https://www.dos.ny.gov/opd/sser/pdf/AllSigns_Islip.pdf, accessed November 27, 2020.
33. *Providence* [RI] *Evening Tribune*, February 1, 1925, 46.
34. Lumsden et al., "A Typhoid Fever Epidemic Caused by Oyster-Borne Infection (1924–1925)," 100.
35. *New York Times*, March 8, 1925, XX 5.
36. *Brooklyn Times Union*, December 5, 1926, 37.
37. Jennifer Brizzi, "Oysters," in *Savoring Gotham: A Food Lover's Companion to New York City*, ed. Andrew F. Smith. (New York: Oxford University Press, 2015), 443.
38. Joseph Mitchell, "Mr. Flood's Party," in Joseph Mitchell, *Up in the Old Hotel* (New York: Vintage, 2008), 419–20.
39. Joseph Mitchell, "The Rivermen," in *Up in the Old Hotel*, 581.
40. George C. Matthiessen, "A Review of Oyster Culture and the Oyster Industry in North America," Woods Hole Oceanographic Institute, August 1971, 3.
41. David M. Dressel and Donald R. Whitaker, "The U.S. Oyster Industry: An Economic Profile for Policy and Regulatory Analysts," January 1983, 29; https://babel.hathitrust.org/cgi/pt?id=uc1.31822031333040&view=1up&seq=4&skin=2021.
42. *Annual Report of the Department of Health of the City of New York*, 1919, 203.
43. Joseph Mitchell, "Mr. Hunter's Grave," in *Up in the Old Hotel*, 522.
44. E. P. Churchill, Jr., "The Oyster and the Oyster Industry of the Atlantic and Gulf Coasts," Bureau of Fisheries Document No. 890 (Washington, DC: Government Printing Office, 1920), 24.
45. *Brooklyn Daily Eagle*, May 28, 1889, 37; "Ockers, Jacob," 207.
46. *Hartford Daily Courant*, February 2, 1930, 6E.
47. *Sunset* 172 (March 1984).
48. *New York Times*, March 8, 1925, XX 5.
49. Matthiessen, "A Review," 10.
50. *Yonkers Herald Statesman*, November 7, 1938, 12.
51. Erin Becker, "Dutch Baymen, Blue Points, And Oyster Crazed New Yorkers," Gotham Center for New York History, June 6, 2019, https://www.gothamcenter.org/blog/dutch-baymen-blue-points-and-oyster-crazed-new-yorkers.
52. Edward Small, "Oysters, Perennial," *Suffolk County News*, May 7, 1998, 15.
53. Matthiessen, "A Review," 9.
54. Daniel Merriman, "Food Shortages and the Sea," *Yale Review*, c. 1950, 433, Joseph Mitchell Papers, Box 54, Folder 5, New York Public Library, New York, NY.
55. Bill Buford, "On the Bay," *The New Yorker* (April 10, 2006): 35.
56. Paul Greenberg, *American Catch: The Fight for Our Local Seafood* (New York: Penguin, 2014), 34–35.

57. Waldman, *Heartbeats in the Muck*, 22.
58. Allan Ross MacDougall, "Oyster Stew Supreme at Grand Central, New York," in Mark Kurlansky, *The Food of a Younger Land* (New York: Penguin, 2009), 84.
59. Joseph Mitchell, *Bottom of the Harbor* Notes, c. 1950, Joseph Michell Papers, Box 30, Folder 7, Manuscripts and Archives Division, New York Public Library, New York, NY.
60. Mitchell, "The Bottom of the Harbor," in *Up in the Old Hotel*, 465, 487.
61. Clyde L. MacKenzie, Jr., *The Fisheries of Raritan Bay* (New Brunswick, NJ: Rutgers University Press, 1992), 237.
62. Grace Lichtenstein, "Fish Scares Trim Sales and Worry Fulton Dealers," *New York Times*, June 18, 1971, 41.
63. Schuyler, *Embattled River*, 5.
64. Barbara Rader, "If It Isn't Fresh, You're Out of Business, Fish Buyer Says," *Newsday*, July 28, 1976, 2A; Robin Shulman, *Eat the City* (New York: Broadway Books, 2012), 250–52.
65. Schuyler, *Embattled River*, 82.
66. Mathiessen, "A Review," 43.
67. Oyster Obsession, "How (and Why) to Read an Oyster Tag," https://oyster-obsession.com/how-to-read-an-oyster-tag/, accessed December 9, 2020.
68. MacDougall, "Oyster Stew Supreme at Grand Central, New York," 85.
69. Jerome Brody, "Welcome to the Grand Central Oyster Bar & Restaurant," in Sandy Ingber, *Grand Central Oyster Bar & Restaurant Complete Seafood Cookbook* (New York: Stewart, Tabori & Chang, 1999), x.
70. Suzanne Hamlin, "On Behalf of the Beautiful Bivalve," *New York Daily News*, "Good Living," January 29, 1986, 3.
71. Sandy Ingber with Roy Finamore, *The Grand Central Oyster Bar & Restaurant Cookbook* (New York: Stewart, Tabori & Chang, 2013), 22.
72. Suzanne Hamlin, "An Oyster for All Seasons," *New York Daily News*, Good Living, March 7, 1990, 3.
73. Mark Kurlansky, *The Big Oyster: History on the Half Shell* (New York: Ballantine, 2006), 247.
74. *New York Times*, September 14, 1959, 34.
75. Tom Schlichter, "Growing Culture: Oyster Farming's Resurgence on Long Island," *Newsday*, December 7, 2019, https://www.newsday.com/long-island/growing-culture-oyster-farming-s-resurgence-on-li-1.39248582.
76. Hamlin, "On Behalf," 7.
77. Mitchell, "The Bottom of the Harbor," 472.

11. BUYERS

1. Sandra L. Oliver, *Saltwater Foodways: New Englanders and Their Food, at Sea and Ashore, in the Nineteenth Century* (Mystic, CT: Mystic Seaport Museum, 1995), 354.
2. Bruce Stutz, "Fish Market Taxonomy," *Natural History* 95 (August 1986): 76.

3. Paul Greenberg, *Four Fish: The Future of the Last Wild Food* (New York: Penguin, 2010), 84.
4. *Scientific American* 37 (July 28, 1877): 55.
5. G. Brown Goode, "The Food Fishes of the United States," in *The Fisheries and Fishery Industries of the United States*, ed. George Brown Goode, Section I: Natural History of Useful Aquatic Animals (Washington, DC: Government Printing Office, 1884), 321.
6. *New York Times*, October 11, 1897, 3.
7. Thomas De Voe, *The Market Assistant* (New York: Hurd and Houghton, 1867), 268.
8. Sasha Issenberg, *The Sushi Economy: Globalization and the Making of a Modern Delicacy* (New York: Gotham Books, 2007), 168.
9. *New York Sun*, July 25, 1909, Section 2, 2.
10. Elizabeth Pillsbury, "An American Bouillabaisse: The Ecology, Politics and Economics of Fishing around New York City, 1870–Present" (PhD diss., Columbia University, 2009), 131–32.
11. Joseph Mitchell, "Dragger Captain," in Joseph Mitchell, *Up in the Old Hotel* (New York: Vintage, 2008), 569.
12. Sandy Ingber with Roy Finamore, *The Grand Central Oyster Bar & Restaurant Cookbook* (New York: Stewart, Tabori & Chang, 2013), 13.
13. Michael and Ariane Betterberry, *On the Town in New York: From 1776 to the Present* (New York: Charles Scribner's Sons, 1973), 171–72.
14. William Grimes, *Appetite City: A Culinary History of New York* (New York: North Point Press, 2009), 133–42; George S. Chappell, *The Restaurants of New York* (New York: Greenberg, 1925), 23.
15. Joseph Mitchell, "A Mess of Clams," in *Up in the Old Hotel*, 312–13.
16. Batterberry and Batterberry, *On the Town in New York*, 141.
17. Barbara Rader, "Where Fish Is Favored by Men Who Know," *Newsday*, December 2, 1966, 1B. That patron is described as "a writer." I like to think it was Joseph Mitchell, but who knows?
18. John H. Matthews, "A History of the Fisheries Industries of New York," *Fishing Gazette* (Annual Review 1929): 70.
19. Ellen Fletcher Rosebrock, "Sweet's Refectory," *Seaport* 15 (Winter 1982): 20.
20. Batterberry and Batterberry, *On the Town in New York*, 324.
21. Charles B. Driscoll, "New York Day by Day," *The* [New London, CT] *Day*, April 26, 1938, 6.
22. Joseph Mitchell, "Up in the Old Hotel," in *Up in the Old Hotel*, 440, 444–45.
23. Mitchell, "Up in the Old Hotel," 445.
24. Joan Cook, "The Fulton Fish Market: More Than a Place to Find Sea Food," *New York Times*, September 5, 1966, R27.
25. "Sloppy Louie's," Joseph Mitchell Papers, Box 72, Folder 10, New York Public Library, New York, NY.
26. John Von Glahn, "Fulton Fish Market N.Y.'s 'Grand Old Lady,'" *Fishing Gazette* (July 1978): 10, in Joseph Mitchell Papers, Box 976.3, No. 1, New York Public Library, New York, NY.

27. Pillsbury, "An American Bouillabaisse," 177–78; *The* [New London, CT] *Day*, February 25, 1950, 16.
28. Oliver, *Saltwater Foodways*, 348.
29. Pillsbury, "An American Bouillabaisse," 176.
30. W. O. Saunders, "Here's a Million For Some Live Fisherman," *Elizabeth City* [NC] *Independent*, December 23, 1927, 1.
31. Russ Symontowne, "Lent Fasting Booms Big N.Y. Fish Mart," *New York Daily News*, March 17, 1936, 37.
32. June Owen, "Food News: Shad Boning Rates as Art," *New York Times*, April 7, 1961, 22.
33. Nan Ickeringill, "Food: The Shad Season," *New York Times*, March 24, 1964, 39.
34. Dana Bowen, "Fulton Market Looks Forward to Bronx Dawns," *New York Times*, March 23, 2005, https://www.nytimes.com/2005/03/23/style/dining/fulton-market-looks-forward-to-bronx-dawns.html?searchResultPosition=1.

    Paul Josephson argues that "Consumers were not attracted by the form of these frozen fillets, and demand for fish products remained low." That was true only of mass-produced, machine-cut fillets that required chemical dipping, not the higher-end fillets coming out of places like the Fulton Fish Market. See Paul Josephson, "The Ocean's Hot Dog: The Development of the Fish Stick," *Technology and Culture* 49 (January 2008): 41–42.
35. Bruce Beck, *The Official Fulton Fish Market Cookbook* (New York: Dutton, 1989), 25–27.
36. William W. Warner, *Distant Water: The Fate of North Atlantic Fishermen* (New York: Penguin, 1983), 41.
37. Barbara Mensch, *The Last Waterfront: The People of South Street* (New York: Freundlich Books, 1985), 62.
38. Dick Ryan, "Countdown on Fulton Street," *New York Daily News*, Sunday News Magazine, October 19, 1975, 27.
39. Oliver, *Saltwater Foodways*, 348–49.
40. Mansel G. Blackford, *Making Seafood Sustainable* (Philadelphia: University of Pennsylvania Press, 2012), 178.
41. Pillsbury, "An American Bouillabaisse," 191–92.
42. Jane Nickerson, "Fish Sticks Soar in Public Favor With New Makers by Dozens in Field," *New York Times*, May 20, 1954, 37.
43. Josephson, "The Ocean's Hot Dog," 48.
44. Mansel G. Blackford, "Fishers, Fishing, and Overfishing: American Experiences in Global Perspective, 1976–2006," *Business History Review* 83 (Summer 2009): 256.
45. Philip Lopate, "Fish Tale: Falling for a Live One," *New York Times*, 5 January 2001, E48.
46. Bowen, "Fulton Market Looks Forward to Bronx Dawns."
47. Molly O'Neill, "A Day Dawns at Fulton," *Newsday*, March 4, 1987, C1.
48. Priscilla Jennings Slanetz, "A History of the Fulton Fish Market," *The Log of Mystic Seaport* 38 (1986): 21. The list from that year included categories for "small freshwater

fish" and "small saltwater fish," but it is hard to imagine that breaking those categories out would have changed the overall number of species much.
49. Jeanne Lesem (United Press International), "Seafood Tradition Changing," *Rome [GA] News-Tribune*, February 26, 1975, 6-B.
50. O'Neill, "A Day Dawns at Fulton," C1.
51. Bernadette Wheeler, "High Tide of Seafood," *Newsday*, March 4, 1987, C3.
52. Alex Dominque (Associated Press), "Eating Squid Is a Real Culinary Adventure," [Washington, PA] *Observer-Reporter*, November 8, 1987, E-6.
53. Trevor Corson, *The Zen of Fish: The Story of Sushi from Samurai to Supermarket* (New York HarperCollins, 2007), 237.
54. "New York City's FULTON FISH MARKET," WPA, Federal Writers Project, NYC Unit, 1940, "Feeding the City," Microfilm Roll 131, New York Municipal Archives, New York, NY.
55. Issenberg, *The Sushi Economy*, xiii, 168, 146.
56. Dan Barber, *The Third Plate: Field Notes on the Future of Food* (New York: Penguin, 2014), 252–53.
57. Corby Kummer, "The Fresh Fish Myth," *New York Magazine* 28 (June 26, 1995–July 3, 1995): 76.
58. Stutz, "Fish Market Taxonomy," 76–77.
59. Karen Bruno, "All in a Day's Catch for N.Y," *Nation's Restaurant News* 22 (March 21, 1988).
60. Barber, *The Third Plate*, 255–57.
61. William W. Warner, "At the Fulton Market," *The Atlantic* 236 (November 1975): 60.

12. THE CULTURE OF THE FULTON FISH MARKET AND ORGANIZED CRIME

1. Matt A.V. Chaban, "Finding Glamour, Not Grit, at South Street Seaport," *New York Times*, January 5, 2015, https://www.nytimes.com/2015/01/06/nyregion/missing-the-grit-of-the-old-south-street-seaport.html.
2. Barbara Mensch, *The Last Waterfront: The People of South Street* (New York: Freundlich Books, 1985), and Barbara G. Mensch, *South Street* (New York: Columbia University Press, 2007).
3. Joseph Mitchell, "Mr. Flood's Party," in Joseph Mitchell, *Up in the Old Hotel* (New York: Vintage, 2008), 425.
4. *South Street Reporter* 2 (March 1968): 1.
5. Suzanne Hamlin, "A Man with Many Fish Tales," *New York Daily News*, Good Living," March 6, 1985, 3. Emphasis in original.
6. Mensch, *South Street*, 101.
7. Philip Lopate, "Introduction," in Mensch, *South Street*, 19.
8. Phillip Lopate, "Fish Tale: Falling for a Live One," *New York Times*, January 5, 2001, E37.

9. Nina Roberts, "Raising Anchor," *New York Times*, December 12, 2004, https://www.nytimes.com/2004/12/12/nyregion/raising-anchor.html?searchResultPosition=1.
10. *Seaport* 26 (Winter 1992): 31.
11. Charlotte Hughes, "A New Day for the Old Fulton Fish Market," *New York Times Magazine*, April 24, 1938, 12.
12. Mensch, *South Street*, 84, 91.
13. Interview with author Kevin Baker in Corinna Montlo and Alex Brook Lynn, "Up At Lou's Fish: The Fulton Fish Market Chronicles," YouTube, accessed February 4, 2021, https://www.youtube.com/watch?v=uWZznLf4IoE&list=FLgcxyFpx4SMxryxYme61xWg&index=1&t=122s&ab_channel=Miss1932, at about 15 minutes.
14. William B. Helmreich, *The New York Nobody Knows: Walking 6000 Miles in the City* (Princeton, NJ: Princeton University Press, 2013), 317.
15. Joseph Mitchell, "Interview With Bob Cantalupo," October 26, 1987, 6, Joseph Mitchell Papers, Box 72, Folder 4, no. 1, Archives and Manuscripts Division, New York Public Library, New York, NY. I have significantly edited the text of Mitchell's notes to improve readability without changing the meaning.
16. *New York Times*, April 6, 1957.
17. Leo Egan, "Official Testifies Stone Heard Lanza Jail Tapes," *New York Times*, May 14, 1957, 28.
18. *Elmira [NY] Advertiser* (Associated Press), May 10, 1957, 15.
19. Robert J. Kelly, *The Upper World and the Underworld: Case Studies of Racketeering and Business Infiltrations in the United States* (New York: Kluwer Academic Publishing, 1999), 109.
20. James B. Jacobs with Coleen Friel and Robert Radick, *Gotham Unbound: How New York Was Liberated from the Grip of Organized Crime* (New York: New York University Press, 1999), 43–45.
21. Selwyn Raab, *Five Families: The Rise, Decline and Resurgence of America's Most Powerful Mafia Empires* (New York: Thomas Dunne, 2016), 585–86.
22. Murray Kempton, "Betterment is the Ruin of Fulton St.," *Newsday*, March 31, 1985, E6.
23. Arnold H. Lubasch, "8 Indicted in Fulton Fish Market Payoffs," *New York Times*, August 14, 1981, B3; D. J. Saunders and Robert Carroll, "On Scent of Fish Mart Crime," *New York Daily News*, October 1, 1982, 40.
24. Arnold H. Lubasch, "2 Brothers Convicted of Fulton Market Extortion," *New York Times*, October 31, 1981, 1, 31.
25. Selwyn Raab, "For Half Century, Tradition Has Foiled Attempts to Clean Up Fulton Fish Market," *New York Times*, April 1, 1995, 22.
26. Arnold H. Lubasch, "U.S. Prosecutors Say Mob Controls Fulton Market," *New York Times*, February 14, 1982, 53.
27. Jacobs, *Gotham Unbound*, 147–49.
28. Arnold H. Lubasch, "Mafia Runs Fulton Fish Market, U.S. Says in Suit to Take Control," *New York Times*, October 16, 1987, B4.
29. *United States v. Local 359*, 705 F. Supp. 894 (S.D.N.Y. 1989).

12. THE CULTURE OF THE MARKET AND ORGANIZED CRIME 265

30. City of New York Department of Investigation, "An Investigation Into Allegations of Criminal Activity at the Fulton Fish Market," April 1992, 4, Joseph Mitchell Papers, Box 102, Folder 8, New York Public Library, New York, NY.
31. City of New York Department of Investigation, 18–19, 21.
32. Randy M. Mastro, "How We'll Clean up the Fulton Market," *New York Daily News*, March 31, 1995, 15C.
33. Jacobs, *Gotham Unbound*, 155.
34. Helene Olen, "N.Y. Crackdown Causes Big Stink at Fulton Fish Market . . .," *Los Angeles Times*, October 20, 1995, https://www.latimes.com/archives/la-xpm-1995-10-20-mn-59079-story.html; Tom Robbins, "New Stink at Fish Market," *New York Daily News*, October 22, 1995, 121.
35. Steven Malanga, "How To Run the Mob Out of Gotham," *City Journal* (Winter 2001), https://www.city-journal.org/html/how-run-mob-out-gotham-12123.html.
36. See for example, Randy Mastro, "We're Winning War on the Mob," *New York Daily News*, September 12, 1997, 47.
37. Lopate, "Introduction," 21.
38. As quoted in Mensch, *South Street*, 68.
39. Mensch, *South Street*, 96.
40. Raab, *Five Families*, 566.
41. Joseph Mitchell, Fulton Fish Market Notes, June 1978, Joseph Mitchell Papers, Box 97, Folder 5.
42. Raab, *Five Families*, 566, 754.
43. Bella English, "Still Coming up Roses at the Fish Market," *New York Daily News*, August 26, 1981, 6.
44. *New York Times*, December 11, 1982, 34; Selwyn Raab, "Crackdown on Mob at Fish Market Brings Chaos," *New York Times*, October 17, 1995, A1.
45. Frank Lombardi, "The Catch of the Year," *New York Daily News*, Weekly News Magazine, December 12, 1982, 7.
46. Jacobs, *Gotham Unbound*, 41.
47. *The* [New London, CT] *Day*, December 21, 1931, 1.
48. Thomas J. Lueck, "Buyers for Restaurants Scramble to Find Fish," *New York Times*, October 17, 1995, B5.
49. Mensch, *South Street*, 126–27.
50. Mensch, *The Last Waterfront*, 61–62.
51. Dan Ackman, "The Big Man in Shrimp," *New York Times*, July 2, 2000, CY4.
52. Kelly, *The Upper World and the Underworld*, 177.
53. Ruth Spear, "Food," in *Avenue* (April 1981): 102, 104 in "FISH MARKET—HISTORIES," South Street Seaport Museum, New York, NY.
54. Leslie Bennetts, "Fulton Market Beset by Crime and Competition," *New York Times*, September 17, 1981, B3.
55. Priscilla Jennings Slanetz, "A History of the Fulton Fish Market," *The Log of Mystic Seaport* 38 (1986): 23.

13. A MUSEUM AND TWO SHOPPING MALLS

1. *New York Sun,* July 16, 1869, 1.
2. Joseph Mitchell, "Mr. Flood's Party," in *Up in the Old Hotel* (New York: Vintage, 2008), 432–33.
3. Stephen Michael Kolman, "We'll Take Manhattan: The Appropriation of Immigrant Space and the Transformation of Urban Geography in New York City, 1925–1975" (PhD diss., University of Wisconsin—Madison, 2002), 6–7.
4. Madigan-Hyland Consulting Engineers, "Fulton Fish Market Study," Department of Public Works, NYC, 1953, I, New York Municipal Archives, New York, NY; Ellen Fletcher, *Walking Around in South Street,* rev. ed. (Stony Creek, CT: Leete's Island Books, 1999), 47.
5. James M. Lindgren, *Preserving South Street Seaport: The Dream and Reality of a New York Urban Renewal District* (New York: New York University Press, 2014), 23. This is the definitive history of the South Street Seaport Museum, the South Street Seaport Corporation, and the development that both these institutions championed.
6. John Von Glahn, "Fulton Fish Market N.Y.'s 'Grand Old Lady,'" *Fishing Gazette* (July 1978): 8. in Mitchell Papers, Box 97, Folder 3.
7. Lindgren, *Preserving South Street Seaport,* 35.
8. Jan Hird Pekorny, "SCHERMERHORN ROW BLOCK—CHRONOLOGY OF EVENTS," South Street Seaport Museum Collections, n.d., 1.
9. Joseph Mitchell, "Tuesday, July 16, 1968," Joseph Mitchell Papers, Box 72, Folder 8, Archives and Manuscripts Division, New York Public Library, New York, NY. Mitchell's notes include many easy-to-follow abbreviations. For clarity's sake, I have expanded them into entire words.
10. Colson Whitehead, *The Colossus of New York* (New York: Anchor Books, 2004), 3–4.
11. John Leland, "Where a Fresh-Caught Fish Is the Best Neighbor," *New York Times,* October 2, 2003, https://www.nytimes.com/2003/10/02/garden/where-a-fresh-caught-fish-is-the-best-neighbor.html?searchResultPosition=10.
12. David Rockefeller to Governor Nelson Rockefeller as quoted in Robin Foster, "Battle of the Port: Memory, Preservation, and Planning in the Creation of the South Street Seaport Museum," *Journal of Urban History* 39 (September 2013): 893.
13. Foster, "Battle of the Port," 892–94.
14. John T. Metzger, "The Failed Promise of a Festival Marketplace: South Street Seaport in Lower Manhattan," *Planning Perspectives* 16 (2001): 29.
15. McCandlish Phillips, "Seaport Museum Urged Downtown," *New York Times,* May 15, 1967, 45.
16. Lindgren, *Preserving South Street Seaport,* 276.
17. H. M. Frankel, "Seaport Project Keys on Heritage," *New York Times,* January 23, 1969, 32. "The Fish Market will come down and be replaced by a facsimile of the famous old market of 1821," Frankel reported in his description of the original Seaport revitalization plan. If you define the "Fish Market" as just the original retail building of 1822, this is precisely what happened.

18. Charles Evans Hughes, "The Water Street That Was," *South Street Reporter* 2 (Summer 1972): 8.
19. Jonathan Barnett, Edward L. Barnes, and James Ulmer and Company, "The South Street Seaport Development Plan," 1974, 2–3, New York Municipal Archives, New York, NY.
20. Lindgren, *Preserving South Street Seaport*, 38.
21. Lindgren, *Preserving South Street Seaport*, 301, n. 9.
22. Dick Ryan, "Countdown on Fulton Street," *New York Daily News*, Sunday News Magazine, October 19, 1975, 34.
23. Bernard Edelman and Sharon Goldstein, untitled Fulton Fish Market article typescript, c. 1975, 19, Joseph Mitchell Papers, Box 102, Folder 7, New York Public Library, New York, NY; *Seaport* 26 (Winter 1992): 41.
24. *Business Week* 2047 (November 23, 1968): 82–83.
25. Ada Louise Huxtable, "A New City Is Emerging Downtown," *New York Times*, March 29, 1970, R4.
26. Barnett et al., "The South Street Seaport Development Plan," 6.
27. Frank J. Prial, "South St. Seaport Seen as Key to Manhattan Landing," *New York Times*, April 17, 1972, 65.
28. Metzger, "The Failed Promise of a Festival Marketplace," 29; Michael Z. Wise, "Modest Endeavors: Reclaiming the Shoreline," in *The New York Waterfront: Evolution and Building Culture of the Port and Harbor*, ed. Kevin Bone (New York: The Monacelli Press, 1997), 243–44.
29. Lindgren, *Preserving South Street Seaport*, 130.
30. McKeown and Franz, Inc., "Seaport Marketplace Environmental Impact Statement," 1981, vi–58, New York Municipal Archives, New York, NY.
31. Owen Moritz, "Seaport Area May Get 60M Rebuilding," *New York Daily News*, September 28, 1979, 7; Metzger, "The Failed Promise of a Festival Marketplace," 34–35, 37.
32. "Lower Manhattan East: Tour Three," in "New York New York '82," 1982, 34, "SOUTH STREET SEAPORT," South Street Seaport Museum, New York, NY.
33. *Architectural Record*, "Down to the Sea in Shops" (January 1984), Reprint, "SOUTH STREET SEAPORT," South Street Seaport Museum,
34. South Street Seaport Museum, "Visit the New Fulton Market," South Street Seaport Museum.
35. *Architectural Record*, "Down to the Sea in Shops."
36. Wise, "Modest Endeavors," 244.
37. David W. Dunlap, "New Look Planned for Pier at South Street Seaport," *New York Times*, June 18, 2008, https://www.nytimes.com/2008/06/18/nyregion/18seaport.html?searchResultPosition=5.
38. Lindgren, *Preserving South Street Seaport*, 281–82.
39. "Chronicling The South Street Seaport's Post-Sandy Decline," Curbed New York, September 5, 2013, https://ny.curbed.com/2013/9/5/10201344/chronicling-the-south-street-seaports-post-sandy-decline.
40. Metzger, "The Failed Promise of a Festival Marketplace," 42.

41. Bernard Stamler, "Rough Sailing for South Street Seaport," *New York Times*, March 29, 1998, https://www.nytimes.com/1998/03/29/nyregion/rough-sailing-for-south-street-seaport.html?searchResultPosition=10.
42. Monte Williams, "Fulton Market Building: Less Is More and Super Is Better," *New York Times*, April 30, 1995, CY 6.
43. Stamler, "Rough Sailing for South Street Seaport."
44. Phillip Lopate, "Introduction," in Barbara G. Mensch, *South Street* (New York: Columbia University Press, 2007), 37.
45. William McKibben, *The New Yorker* (July 22, 1985), https://www.newyorker.com/magazine/1985/07/22/fish-market-art.
46. Deirdre Carmody, "Rejuvenated Seaport Is Due to Open July 28," *New York Times*, July 15, 1983, B3.
47. Barbara G. Mensch, *South Street* (New York: Columbia University Press, 2007), 91.
48. Laurie Johnston, "Plan for South Street Market Leaves a Wake Of Dissension," *New York Times*, November 28, 1979, B20.
49. McKeown and Franz, Inc., "Seaport Marketplace Environmental Impact Statement," vi–34.
50. Stamler, "Rough Sailing for South Street Seaport."
51. Lindgren, *Preserving South Street Seaport*, 35.
52. Paul Goldberger, "At Seaport, Old New York With a New Look," *New York Times*, July 29, 1983, C17.
53. "Lower Manhattan East: Tour Three," 32.
54. Stamler, "Rough Sailing for South Street Seaport."
55. Didi Moore, "South Street Seaport Sails into the 21st Century," *New York Daily News*, July 2, 1981, 58.
56. Marianne Baker, "Where's There's Water, There Must Be Good Fish Nearby: Some Great Places to Find It and Eat It," *The Villager*, July 2, 1981, 19.
57. Lindgren, *Preserving South Street Seaport*, 193.

14. RELOCATION

1. Dan Barry, "A Last Whiff of Fulton's Fish, Bringing a Tear," *New York Times*, July 10, 2005, https://www.nytimes.com/2005/07/10/nyregion/a-last-whiff-of-fultons-fish-bringing-a-tear.html?searchResultPosition=1.
2. *New York World*, June 8, 1879, 2.
3. Madigan-Hyland Consulting Engineers, "Fulton Fish Market Study," Department of Public Works, NYC, 1953, v, New York Municipal Archives, New York, NY.
4. Phillip Lopate, "Introduction," in Barbara G. Mensch, *South Street* (New York: Columbia University Press, 2007), 30.
5. Charles H. Brown, "Downtown Enters a New Era," *New York Times Magazine*, January 31, 1960, 70.

6. Anthony Robins, *The World Trade Center* (Englewood, FL: Pineapple Press, 1987), 14–15.
7. Robert A.M. Stern, Thomas Mellins, and David Fishman, *New York: 1960: Architecture and Urbanism Between the Second World War and the Bicentennial* (New York: The Monacelli Press, 1995), 198, 200.
8. Robins, *The World Trade Center*, 12, 17; Gilbert Millstein, "Restless Ports for the City's Food," *New York Times Magazine*, April 24, 1960, 62.
9. *New York Daily News*, August 19, 1955, 7.
10. Anne Babette Audant, "From Public Market to *La Marqueta*: Shaping Spaces and Subjects of Food Distribution in New York City, 1930 to 2012" (PhD diss., City University of New York, 2013), 149–50; Phillip Lopate, *Waterfront: A Walk Around Manhattan* (New York: Anchor, 2004), 235.
11. Robin Shulman, *Eat the City* (New York: Broadway, 2012), 126.
12. John Toscano, "Fulton Fish Mart Going Upstream To a New Perch," *New York Daily News*, August 21, 1974, 7.
13. Edwards and Kelcey, "Wholesale Fish Market for New York City: Survey, Study, Report," New York City Department of Public Works, 1963, 3. New York Municipal Archives, New York, NY. *Commercial Fisheries Review* 26 (November 1964): 41.
14. Carol Willis, ed., *The Lower Manhattan Plan: The 1966 Vision for Downtown New York* (New York: Princeton Architectural Press, 2002), Lower Manhattan Plan 1.
15. Bernard Edelman and Sharon Goldstein, untitled Fulton Fish Market article typescript, c. 1975, 2–3, Joseph Mitchell Papers, Box 102, Folder 7, New York Public Library, New York, NY.
16. Bruce Beck, *The Official Fulton Fish Market Cookbook* (New York: Dutton, 1989), 15.
17. Edelman and Goldstein, 4–7.
18. Peter Kihss, "Work Starts at Hunts Point in October On Long Planned Fulton Fish Market," *New York Times*, August 21, 1974, 43.
19. Priscilla Jennings Slanetz, "A History of the Fulton Fish Market," *The Log of Mystic Seaport* 38 (1986): 24.
20. Frank McKeown, "Fish Mart Will Swim to Bronx," *New York Daily News*, April 16, 1969, 28.
21. Lopate, "Introduction," 32.
22. Richard Baiter, "Lower Manhattan Waterfront" (New York: Office of Lower Manhattan Development, 1975), 34.
23. Dick Ryan, "Countdown on Fulton Street," *New York Daily News*, Sunday News Magazine, October 19, 1975, 28.
24. Walter Chambers, *Labor Unions and the Public* (New York: Coward-McCann, c. 1936), 150.
25. Ryan, "Countdown on Fulton Street," 28.
26. Joseph Mitchell, Fulton Fish Market Notes, June 1978, Joseph Mitchell Papers, Box 97, Folder 5, Archives and Manuscripts Division, New York Public Library, New York, NY.

27. John L. Hess, "Fish-Market Move to Bronx Impeded," *New York Times*, July 26, 1974, 36.
28. Ryan, "Countdown on Fulton Street," 31.
29. Edwards and Kelcey, "Wholesale Fish Market for New York City: Survey, Study, Report," New York City Department of Public Works, 1963, 4, New York Municipal Archives, New York, NY.
30. Corby Kummer, "The Fresh Fish Myth," *New York Magazine* 28 (June 26, 1995–July 3, 1995): 74.
31. Edelman and Goldstein, 12–13.
32. Lopate, "Introduction," 34.
33. Thomas J. Lueck, "City Proceeds With Moving Fulton Fish Market to Bronx," *New York Times*, February 15, 2001, B3.
34. Corinna Montlo and Alex Brook Lynn, "Up At Lou's Fish: The Fulton Fish Market Chronicles," YouTube, accessed February 4, 2001, https://www.youtube.com/watch?v=uWZznLf4IoE&list=FLgcxyFpx4SMxryxYme61xWg&index=1&t=122s&ab_channel=Miss1932, at about 51 minutes.
35. Sasha Issenberg, *The Sushi Economy: Globalization and the Making of a Modern Delicacy* (New York: Gotham Books, 2007), 27–28, 212.
36. Kummer, "The Fresh Fish Myth," 74.
37. James Beard, *James Beard's Fish Cookery* (New York: Warner Books, 1967), 11. The 1967 edition is a reprint. When Beard revised the book in 1976, he kept this line.
38. Kummer, "The Fresh Fish Myth," 74.
39. Paul Greenberg, *American Catch: The Fight for Our Local Seafood* (New York: Penguin, 2014), 9.
40. Dan Barber, *The Third Plate: Field Notes on the Future of Food* (New York: Penguin, 2014), 218.
41. Anthony Bourdain, "Don't Eat Before Reading This," *The New Yorker*, April 12, 1999, https://www.newyorker.com/magazine/1999/04/19/dont-eat-before-reading-this.
42. Anthony Bourdain, "The Fish-on-Monday Thing," in *Medium Raw: A Bloody Valentine to the World of Food and the People Who Cook* (New York: Ecco, 2010): 268.
43. *The New Yorker* (November 18, 1974): 47.
44. Jeanne Lesem (United Press International), "Seafood Tradition Changing," *Rome [GA] News-Tribune*, February 26, 1975, 6-B.
45. Kathryn Graddy, "Markets: The Fulton Fish Market," *Journal of Economic Perspectives* 20 (Spring 2006): 208–09.
46. John Waldman, *Heartbeats in the Muck: The History, Sea Life and Environment of New York Harbor*, rev. ed. (New York: Empire State Editions, 2013), 42.
47. Hess, "Fish-Market Move to Bronx Impeded," 36.
48. Dana Bowen, "Fulton Market Looks Forward to Bronx Dawns," *New York Times*, March 23, 2005, F1.
49. Francis J. Duffy and William H. Miller, *The New York Harbor Book* (Falmouth, ME: TBW Books, 1986), 154-55.
50. Beck, *The Official Fulton Fish Market Cookbook*, 23.

51. Selwyn Raab, "Crackdown on Mob at Fish Market Brings Chaos," *New York Times*, October 17, 1995, B7.
52. Martin Mbugua and Lisa L. Colangelo, "Fish Market Move Snarled," *New York Daily News*, February 15, 2001, 17.
53. Charlie Leduff, "The Fish Return to Fulton Market (Along With the Mongers)," *New York Times*, October 16, 2001, D5; "Charlie Leduff, "Work Begins on Fish Market," *New York Times*, November 28, 2001, D4.
54. Lueck, "City Proceeds With Moving Fulton Fish Market to Bronx," B3.
55. Kevin McCarty, "Market Fresh," *Civil Engineering* 76 (June 2006): 63.
56. Bob Kappstatter, "Fish Mart's Debut Delayed," *New York Daily News*, September 23, 2005, 2CN.
57. Helen Peterson, "Fulton Fish Market OKs Deal to Open Shop in Bx.," *New York Daily News*, November 8, 2005, 3.
58. Lopate, *Waterfront*, 243.
59. As transcribed from Montlo and Lynn, "Up At Lou's Fish," at about 1 hour, 18 minutes in.

CONCLUSION

1. Nina Lalli, "On the Job: The New Fulton Fish Market," *The Village Voice*, October 10, 2007, https://www.villagevoice.com/2007/10/10/on-the-job-the-new-fulton-fish-market/.
2. Ben Sargent, "Catch It, Cook It & Eat It—Fulton Fish Market New vs Old," YouTube, May 28, 2010, https://www.youtube.com/watch?v=0RzCxF4lUME&ab_channel=BenSargent.
3. Jordana Rothman, "Chefs Hit the Fulton Fish Market, Pop Uni and Bacon-Buttered Crab Meat Until 5 a.m.," *Food & Wine*, May 23, 2017, https://www.foodandwine.com/seafood/chefs-hit-fulton-fish-market-pop-uni-and-bacon-buttered-crab-meat-until-5-am.
4. Larry Sutton, "Catch of the Day," *New York Daily News*, December 3, 1995, 46.
5. New Fulton Fish Market, "About," http://www.newfultonfishmarket.com/about.html; Andrew Jacobs, "Little Love Lost Among Fishmongers for Fulton Location," *New York Times*, November 15, 2005, https://www.nytimes.com/2005/11/15/nyregion/little-love-lost-among-fishmongers-for-fulton-location.html?searchResultPosition=12; New Fulton Fish Market, "New Fulton Fish Market Rules And Regulations," accessed May 18, 2021, http://www.newfultonfishmarket.com/rules.html.
6. Manny Fernandez, "The New Fulton Fish Market," *New York Times* video, July 12, 2006, https://www.nytimes.com/video/nyregion/1194817112350/the-new-fulton-fish-market.html?searchResultPosition=1.
7. Manny Fernandez, "In Sleek Fish Market, Missing Romance of the Old Ways," *New York Times*, July 13, 2006, https://www.nytimes.com/2006/07/13/nyregion/13journal.html?searchResultPosition=70.

8. Dan Barry, "Just Scents and Memories, Wafting on the Breeze," *New York Times*, August 5, 2006, https://www.nytimes.com/2006/08/05/nyregion/just-scents-and-memories-wafting-on-the-breeze.html?searchResultPosition=3.
9. John Freeman Gill, "A Slice of the Fulton Fish Market Gets a New Life," *New York Times*, February 28, 2020, https://www.nytimes.com/2020/02/28/realestate/a-slice-of-the-fulton-fish-market-gets-a-new-life.html.
10. Jane Margolies, "5 Sites That Show How Much Lower Manhattan Has Changed," *New York Times*, January 24, 2020, https://www.nytimes.com/2020/01/24/nyregion/downtown-manhattan-history.html?searchResultPosition=5.
11. "Chronicling The South Street Seaport's Post-Sandy Decline," Curbed New York, September 5, 2013, https://ny.curbed.com/2013/9/5/10201344/chronicling-the-south-street-seaports-post-sandy-decline. "Bad Guy Joe," "Fulton Fish Market—LTV Squad," October 20, 2021, https://ltvsquad.com/2021/10/20/fulton-fish-market/.
12. Dick Ryan, "Countdown on Fulton Street," *New York Daily News*, Sunday News Magazine, October 19, 1975, 34.
13. National Oceanic and Atmospheric Administration, "Laws & Policies," accessed August 2, 2021, https://www.fisheries.noaa.gov/topic/laws-policies. Congress made a few other changes when the SFA was reauthorized in 2007.
14. "Anthony Bourdain: It's okay to eat fish on Mondays now," *Tech Insider*, accessed January 12, 2021, https://www.youtube.com/watch?v=KcPpJhOhdnQ&ab_channel=TechInsider.
15. "Vinny Milburn Shares the Story Behind Greenpoint Fish & Lobster Co.," Diced, December 28, 2019, https://ice.edu/blog/vinny-milburn-greenpoint-fish-lobster-co
16. Nicki Holymard, "Q&A: Fighting the frozen-at-sea stigma," *Seafood Source*, April 17, 2011, https://www.seafoodsource.com/features/q-a-fighting-the-frozen-at-sea-stigma. There are now multiple ways to flash-freeze fish, depending upon the nature of the product. From the standpoint of the history of refrigeration, they are all very interesting. However, from the standpoint of the history of food, these improvements have less significance: except sushi lovers, very few people have been able to tell the difference between fresh and frozen fish since the 1950s. See Postelsia Team, "Take Another look at the Freezer Case," James Beard Foundation blog, October 4, 2019, https://www.jamesbeard.org/blog/take-another-look-at-the-freezer-case.
17. Stefania Vannuccini, "Shark Utilization, Marketing and Trade," FAO Technical Fisheries Paper 389 (Rome: United Nations Food and Agriculture Organization, 1999), 127–28, 91, 133.
18. Carl Safina, *Song for the Blue Ocean* (New York: Holt, 1997), 402.
19. Meg Wilcox, "As Coronavirus Disrupts Seafood Supply Chains, Struggling Fishermen Seek Other Markets," Civil Eats, April 14, 2020, https://civileats.com/2020/04/14/as-coronavirus-disrupts-seafood-supply-chains-struggling-fishermen-seek-other-markets/.
20. Sarah Halzack, "Something Fishy Is Going On in American Kitchens," Bloomberg, January 23, 2021, https://www.bloomberg.com/opinion/articles/2021-01-23/seafood-supermarket-sales-skyrocket-in-pandemic.

21. Halzack, "Something Fishy Is Going On in American Kitchens."
22. Christine Blank, "Landmark New York City Wholesaler Struggling to Survive," SeafoodSource, July 8, 2020, https://www.seafoodsource.com/news/foodservice-retail/landmark-new-york-city-wholesaler-struggling-to-survive.
23. Lizzie Widdicombe, "The Last Robot-Proof Job in America?," *The New Yorker*, August 1, 2019, https://www.newyorker.com/culture/culture-desk/the-last-robot-proof-job-in-america.
24. Trevor Corson, "A Tale of Three Tuna," *Conservation*, August 1, 2008, https://www.conservationmagazine.org/2008/08/a-tale-of-three-tuna/.
25. Charles Clover, *The End of the Line: How Overfishing Is Changing the World and What We Eat* (Berkeley: University of California Press, 2006), 18.
26. Ian Urbina, "The Smell of Money," *The New Yorker*, March 8, 2021, https://www.newyorker.com/magazine/2021/03/08/fish-farming-is-feeding-the-globe-whats-the-cost-for-locals.
27. New York Department of Environmental Conservation, "American Shad," accessed May 15, 2021, https://www.dec.ny.gov/animals/62510.html.
28. Safina, *Song for the Blue Ocean*, 43–44.
29. Paul Greenberg, *American Catch: The Fight for Our Local Seafood* (New York: Penguin, 2014), 11–12.
30. Governor Andrew M. Cuomo, "Governor Cuomo Signs Legislation to Prohibit the Sale and Possession of Shark Fins," December 25, 2012, https://www.governor.ny.gov/news/governor-cuomo-signs-legislation-prohibit-sale-and-possession-shark-fins.
31. Monterey Bay Aquarium Seafood Watch, https://www.seafoodwatch.org/.
32. On fish fraud, see Jonathan Rees, *Food Adulteration and Food Fraud* (London: Reaktion Books, 2020), 68–72.
33. Helen Tangires, *Public Markets and Civic Culture in Nineteenth-Century America* (Baltimore, MD: Johns Hopkins University Press, 2003), xvi.
34. Mansel G. Blackford, *Making Seafood Sustainable* (Philadelphia: University of Pennsylvania Press, 2012), 196–97.
35. Clover, *The End of the Line*, 38.

A NOTE ON SOURCES

1. Thomas Kunkel, *Man in Profile: Joseph Mitchell of The New Yorker* (New York: Random House, 2015), 157.
2. Philip Lopate, "Introduction," in Barbara G. Mensch, *South Street* (New York Columbia University Press, 2007), 25.
3. Lopate, "Introduction," 25.
4. Yes, a regular old cookbook begins with a very serious and scholarly history of the Fulton Fish Market.
5. Theodore Bestor, *Tsukiji: The Fish Market at the Center of the World* (Berkeley: University of California Press, 2004), 13.

INDEX

airplanes, xxii
Alaskan pollock, 159
alewife (menhaden), xv, 76
American Society for the Prevention of Cruelty to Animals (ASPCA), 98
Amos G. Chesebro Company, 58
angler fish, 148
Atlantic Coast Fisheries Company, 27

Baird, Spencer, 85–86
Baltimore, Maryland, xiv, 66, 95, 97, 177, 187
Barbee, Alexander, 89–90
Barnum, P. T., 98
Bay State Fishing Company, 155
Beame, Abraham, 198
Beard, James, 203
beef, ix–x, xxii, 21, 43, 114
Beekman, William, 11
Beekman Slip, 18
Bergh, Henry, 98–99
Birdseye, Clarence, 108, 156, 253n18
Bishop, Kronimus & Pannen, 58
Black Ball Line, 18
blackened redfish, 161
Blackford & Company, 58–59, 70

Blackford, Eugene G., 58–63, 83–84, 180; on relocating the market, 194; on terrapin, 89, 97–98
Bloomberg, Michael, 208
bluefin tuna (horse mackerel), 148–49, 160, 218
bluefish, 2, 37, 39, 78
Bluepoints Oyster Company, 139, 259n32
Bolster, W. Jeffrey, 82
Boston Committee of Markets, 121–22
Boston, Massachusetts, 32–33, 82, 155, 182; as competition for Fulton Fish Market, 175, 196; as landing place for Fulton Market fish, 36–37, 66, 108, 112; as market for fish, 113, 204
Bourdain, Anthony, 203–04, 216
Brooklyn Bridge, xi, 119, 214
Bureau of Chemistry (U.S. Department of Agriculture), 137, 139
Burnet and Kenney Continental Company, 27
Burroughs, John, 86
butchers, 12, 16, 19–22, 24

Caleb Haley and Company, 58
Cambridge, Massachusetts, 32

Camden, Maine, 106
Campbelltown, New Brunswick (Canada), 39
Cantalupo, Bob, 167, 180
Cantalupo, Joe, 180
Carmine's Bar and Grill, 173
Carnegie, Dale, 103
carp, 125
Carter, Joseph H., 205
Castelli, Francesco, 100–101
Catherine Market, 9
Chesebro Brothers, 27
Chicago, Illinois, 67, 76, 112, 188; comparisons with meatpacking in, x, 43, 114, 155, 195; oysters in, 139, 142
cod, x, 2, 31, 33, 70; types of, 37
coffee-and-cake shops, 24
Cold Spring Harbor, New York, 59
cold storage, xxi, 31, 67, 88, 109–11
Collins Line, 18
Comstock & Kingsland, 58
Consolidated Edison, 125
Convention on International Trade in Endangered Species of Wild Fauna and Flora, 101
conservation laws, 75, 90; federal, 85, 87, 159, 215, 221; local, 221
Cooper, James Fenimore, 4
Cortelyou, Peter, 30
Costa Di Mare, ix
COVID-19, 217, 229
crabs, 9, 64, 163, 215
Crain, C. Y., 128
Cranston, Stanley, 132
crawfish, 9

Delmonico's, 152
De Voe, Thomas Farrington, 7, 34, 40, 42, 235–36n23
Dewey, Thomas E., 167
D. Haley and Company, 27, 58
D. Manwaring and Co., 64
Dickens, Charles, 55–56

Dinkins, David, 171
dogfish, 219
dories, 36–37, 47, 76, 78
Dorlan's Tavern, 213
Dorlon's, 53–54, 57
Downing, Thomas, 55–56

East River, 63, 88, 129, 207, 209; boat traffic on, 123; current of, 194; docks, 18, 22, 113, 120; as location for fish market, xi, 10, 14; oyster barges on, 138, 140; water from, 92
Edison, Thomas, 118
Eldred & Haley, 58
ethnic groups, xvi–xvii, 74, 159, 166
Erie Canal, 22–23, 30, 32

FDR Drive, 132, 213
Feeney & Co., xiv
fish: early attempts at freezing, xxi, 31, 105–08, 115; flash freezing of, 108–09, 272n17; freshness of, 104–05; frozen, 162, 201–04; preservation of, 5–6, 31; unloading of, 16; variety of at Fulton Market/Fulton Fish Market, 26, 159, 249n42. *See also names of individual species*
fish cars, 9, 10, 21
Fisher, M.F.K., 57
Fishery Council, 104, 173
fish fillets, 155–58, 161, 262n34
fish fraud, 70
fish hatcheries, 86–87
fish sticks, 157–58, 161
fish wells, 34
flounder, 2
Fly Market, 10–11
Food and Drug Administration, 145
Fraunces Tavern, 163
"Frozen at Sea" (FAS), 216
Fulton Ferry, 18, 120, 125
Fulton Ferry Hotel, 133
Fulton Fish Market Museum, 60–62

Fulton Fishmongers Association, 28–29, 124, 138, 174
Fulton Market/Fulton Fish Market: freshwater fish market, 125; gender at, 165–66; names for, 233n5; original retail building (1822), 12–14, 15, 62, 237n48, 237n60; retail mall (1983), 187–89, 192–93; retail/wholesale building (1869), 28–29, 62; rules of, 19; "Tin Building" (1907), 124, 132, 212; smell of, 179–80; wholesale building (1847), 16, 25; wholesale building (1909), 125–26, 129, 131; wholesale building (1939), 131–32, 212; wholesale/retail building (1883), 63, 124, 180; wholesale shed (1831), 15

General Motors, 135
Genovese crime family. *See organized crime*
George Still Company, 152
Gill, Theodore, 26
Giuliani, Rudolph, 170–72, 174, 207
Gloucester, Massachusetts, xiv, 31, 36–38, 112–13; fish sticks from, 160, 162
golden king clip, 159
Goode, George Brown, 76, 83, 149
Grand Central Oyster Bar, 144–46, 151, 204, 212
green turtle, xvi, 64, 88; catching of, 94–95; conservation of, 101–02; cooking of, 92–92; cruelty against, 98–101; taste of, 91–91; vs. terrapin, 90
Griesa, Thomas, 170–71
grocery stores, 26, 150, 163, 205
grouper, xvii, 6, 161

haddock, 70, 125
halibut, 78, 112
Haley, Caleb, 27
Hall, Harry, 120
Helmreich, William B., 167
herring, x
Hoboken Turtle Club, 94, 96
Howard Hughes Corporation, 188, 212

Hornaday, William Temple, 96
Hudson, Henry, 9
Hudson River, 2, 123, 126, 196; ice from, 32, 105, 110; pollution in, 135, 144; shad from, 40–43, 86
Hunter, George, 142
Hunts Point (Bronx), xi, 163, 198–99, 205, 208, 214

ice, x, 30–34, 63, 104–06
ice crushing machine, 63, 106
"Ice Trust," 105
Ingber, Sandy, 146, 212
Inter City Fish Company, 122

Jamaica Bay, 135, 137
J. C. Comstock & Co., 58
Johns, Jasper, 132
Jones, Emeline, 95

Kalm, Peter, 1
Kingsland & Comstock, 58

La Guardia, Fiorello, 72, 130, 132, 212
Lanza, Joseph "Socks," 127–29, 167–68, 173, 256n40
Laro Maintenance Corporation, 172, 208
Le Bernardin, 162–63
Le Coze, Gilbert, 162
Leland, Warren, 64, 66
Lenape, 1–2
Lindsay, John, 186
lobster, 9, 64, 78
Lopate, Philip, 158, 166, 172, 189
Lower Manhattan Plan, 197

Mackay, Charles, 56
mafia. *See organized crime*
Magnuson-Stevens Fisheries Act, 215
mahi mahi, 159
Manhattan Bridge, 119
Manhattan Oyster Bar and Chop House, 151–52

Marblehead, Massachusetts, 31
Martinique, 32
Masuci, Alphonso, 148–49
menhaden, 75–78. *See also alewife*
Mensch, Barbara, 164, 166, 173, 175, 191
Metropolitan Hotel, 64
Meyer, Danny, 146–47
Miller, Samuel B., 27
Mitchell, Joseph, xvii–xix, 140, 142, 173, 180, 213; as source, 227, 230; "The Bottom of the Harbor," 144, 147; on Brooklyn Bridge and Fulton Market, 119; on bycatch, 151; on neighborhood surrounding Fulton Fish Market, 181; "Old Mr. Flood," xix, 46, 140, 164, 234n22; on Sloppy Louie's, 153–55; on the smell of Fulton Fish Market, 179–80; on terrapin, 97–98, 100; titles of essays, 234n18; "Up in the Old Hotel" (essay), 153
Monterey Bay Aquarium, 220
Moon, George T., 63
moonfish, 148, 159
Moore and Company, 100
Morino, Leah, 193
Morino, Louis, 153–93
mullet, 6, 204

Nash & Crook's Park Row, 94
New Bedford, Massachusetts, 156, 175, 204, 206
New England Fish Company, 122
New Fulton Fish Market, 207, 209–12, 216, 218
New Orleans, Louisiana, 95
New York Aquarium, 62
New York City Board of Health, 138
New York City Department of Markets, 197
New York City Health Department, 73, 140
New York Harbor, 1–2, 30, 47, 134, 144
New York State Hatchery Station, 59
New York Supreme Court, 13

Ockers, Jacob, 139, 142, 259n32
O'Donnell, James, 14
organized crime, 127–29, 167–77, 208; as a tax, 176
Oswego County, New York, 39
oyster bed, 2
oysters, x, 9, 25, 44–57, 141–44, 215; barges, 54–55, 141; Bluepoint, 45; brands of, 45–46; gathering of, 1–2; floating, 49, 137–38, 140; harvesting of, 48–49; raising of, 48; shucking, 47, 53; species of, 45; stands, 52–53, 55

Packer and Haley, 27
Parker, Sarah Jessica, 193
parrot fish (black-barred hogfish), 60, 161
Patchogue, New York, 45
Peck Slip, 11, 18, 125, 175, 213
peddlers, xi, 72–73
perch, 125
Philadelphia, Pennsylvania, 92, 113, 177
Piper, Enoch, 106
pompano, xvii
produce, 23
Prudhomme, Paul, 161

Raab, Selwyn, 173
Rauam, Naima, 190
Rauschenberg, Robert, 132
Rector's, 151
red snapper, xvii, 31, 59, 161, 204
restaurants. *See also names of individual restaurants*
retail fish stores, xxi, 71, 148–49
Rikers Island, 209
Roberts and Graham, 27
Rochester, New York, 22
Rockefeller, David, 133, 183–84, 186, 195–96
Rockefeller, John D., 134
Rockefeller, Nelson, 183–84
Rockland County, New York, 32
Romano, Carmine, 169–71, 174, 198–200
Romano, Peter, 169–71

Romano, Vincent, 169–70
Rorer, Sarah Tyson, 88
Rouse Corporation, 186–88, 191–93

salmon (Atlantic), x, 31, 35, 39–40
salmon (smoked), 64
salmon (Pacific), 39, 215
salt cod, 33, 64
Samuel Z. Chesebro company, 58
Sandusky, Ohio, 107
Sandy Ground (Staten Island), 47, 142
Sandy Hook, New Jersey, 2
Sardi's, 163
S. B. Miller and Company, 27
Schermerhorn Row, 11, 118, 122, 152, 181, 193
schooner, 35
sea bass, 30, 79
Seabury, Samuel, 128
Seabury Committee, 128
seafood. *See names of individual species*
Seafood Watch, 220
Seaman's Church Institute, 126
sea scallops, xvii
Seymour, Jr., Whitney, 183
shad, xv, 40–43, 86–87, 219
shark, 62, 217, 219, 222
Shinnecock, New York, 177
Shopsin, William C., 184
Sloppy Louie's, xvii, 152–55, 193
"smack" (boat), 34, 36, 76, 241n22
smelts, 64, 70
Smith, Adam, 5–6
Smith, Alfred E., xiv–xv, 67, 70, 81, 132
Smith, Sarah, 11
Smitty's Fillet House, 156, 177
South Street Seaport, xxi, 124, 18, 186–93, 209, 266n17
South Street Seaport Museum, xix, 181–86, 229
Spanish mackerel, 78
squid (calamari), 159
Standard Oil, 134

Stanford, Peter, 181, 189
Statue of Liberty, 214
Still, George M., 46, 140
Stonington, Connecticut, 27
striped bass, 2, 64
Sustainable Fisheries Act, 215
Sweet's, 152–53, 193
Swift, Gustavus, x
Syracuse, New York, 22

Tarrytown, New York, 135
telephone, 67, 114
terrapin (diamondback), xvi, 88–91, 95–98, 101–02, 252n52
Thoreau, Henry David, 39
transportation (improvements in), 23, 112–14
trawlers, 115–16, 205, 219
trawling: with long lines, 36–38, 78; with nets, 79–82, 220
Trillin, Calvin, ix
trout, 2, 31, 64, 83
trout show (at Fulton Market), 83–85
Troy, New York, 2
Tsukiji Fish Market (Tokyo), 211, 230
tub trawling, 78
Tudor, Frederic, 32
tuna, 6
turtle soup, 90, 95, 252n52
Twombly, Cy, 132
typhoid fever, 135–40

University of Maryland, 97
United Seafood Workers' Union Local 359, 127
United States Fish Commission, 77, 85

Van Der Donck, Anders, 2
Vongerichten, Jean-Georges, 162
Von Glahn, John, 173

Wagner, Robert, 197
Walton, Izaak, 4

Washington, D.C., 92, 139–40
Washington Market, 196
weakfish, 78
West Sayville, New York, 139
Wilkisson, Frank W., 103

Williamsburg Bridge, 119
Wohl, Frank W., 171
wolf fish, 151
World Trade Center, 195–96, 207, 214

ARTS AND TRADITIONS OF THE TABLE:
PERSPECTIVES ON CULINARY HISTORY

Albert Sonnenfeld, Series Editor

Salt: Grain of Life, Pierre Laszlo, translated by Mary Beth Mader

Culture of the Fork, Giovanni Rebora, translated by Albert Sonnenfeld

French Gastronomy: The History and Geography of a Passion, Jean-Robert Pitte, translated by Jody Gladding

Pasta: The Story of a Universal Food, Silvano Serventi and Françoise Sabban, translated by Antony Shugar

Slow Food: The Case for Taste, Carlo Petrini, translated by William McCuaig

Italian Cuisine: A Cultural History, Alberto Capatti and Massimo Montanari, translated by Áine O'Healy

British Food: An Extraordinary Thousand Years of History, Colin Spencer

A Revolution in Eating: How the Quest for Food Shaped America, James E. McWilliams

Sacred Cow, Mad Cow: A History of Food Fears, Madeleine Ferrières, translated by Jody Gladding

Molecular Gastronomy: Exploring the Science of Flavor, Hervé This, translated by M. B. DeBevoise

Food Is Culture, Massimo Montanari, translated by Albert Sonnenfeld

Kitchen Mysteries: Revealing the Science of Cooking, Hervé This, translated by Jody Gladding

Hog and Hominy: Soul Food from Africa to America, Frederick Douglass Opie

Gastropolis: Food and New York City, edited by Annie Hauck-Lawson and Jonathan Deutsch

Building a Meal: From Molecular Gastronomy to Culinary Constructivism, Hervé This, translated by M. B. DeBevoise

Eating History: Thirty Turning Points in the Making of American Cuisine, Andrew F. Smith

The Science of the Oven, Hervé This, translated by Jody Gladding

Pomodoro! A History of the Tomato in Italy, David Gentilcore

Cheese, Pears, and History in a Proverb, Massimo Montanari, translated by Beth Archer Brombert

Food and Faith in Christian Culture, edited by Ken Albala and Trudy Eden

The Kitchen as Laboratory: Reflections on the Science of Food and Cooking, edited by César Vega, Job Ubbink, and Erik van der Linden

Creamy and Crunchy: An Informal History of Peanut Butter, the All-American Food, Jon Krampner

Let the Meatballs Rest: And Other Stories About Food and Culture, Massimo Montanari, translated by Beth Archer Brombert

The Secret Financial Life of Food: From Commodities Markets to Supermarkets, Kara Newman

Drinking History: Fifteen Turning Points in the Making of American Beverages, Andrew F. Smith

Italian Identity in the Kitchen, or Food and the Nation, Massimo Montanari, translated by Beth Archer Brombert

Fashioning Appetite: Restaurants and the Making of Modern Identity, Joanne Finkelstein

The Land of the Five Flavors: A Cultural History of Chinese Cuisine, Thomas O. Höllmann, translated by Karen Margolis

The Insect Cookbook: Food for a Sustainable Planet, Arnold van Huis, Henk van Gurp, and Marcel Dicke, translated by Françoise Takken-Kaminker and Diane Blumenfeld-Schaap

Religion, Food, and Eating in North America, edited by Benjamin E. Zeller, Marie W. Dallam, Reid L. Neilson, and Nora L. Rubel

Umami: Unlocking the Secrets of the Fifth Taste, Ole G. Mouritsen and Klavs Styrbæk, translated by Mariela Johansen and designed by Jonas Drotner Mouritsen

The Winemaker's Hand: Conversations on Talent, Technique, and Terroir, Natalie Berkowitz

Chop Suey, USA: The Story of Chinese Food in America, Yong Chen

Note-by-Note Cooking: The Future of Food, Hervé This, translated by M. B. DeBevoise

Medieval Flavors: Food, Cooking, and the Table, Massimo Montanari, translated by Beth Archer Brombert

Another Person's Poison: A History of Food Allergy, Matthew Smith

Taste as Experience: The Philosophy and Aesthetics of Food, Nicola Perullo

Kosher USA: How Coke Became Kosher and Other Tales of Modern Food, Roger Horowitz

Chow Chop Suey: Food and the Chinese American Journey, Anne Mendelson

Mouthfeel: How Texture Makes Taste, Ole G. Mouritsen and Klavs Styrbæk, translated by Mariela Johansen

Garden Variety: The American Tomato from Corporate to Heirloom, John Hoenig

Cook, Taste, Learn: How the Evolution of Science Transformed the Art of Cooking, Guy Crosby

Meals Matter: A Radical Economics Through Gastronomy, Michael Symons

The Chile Pepper in China: A Cultural Biography, Brian R. Dott

The Terroir of Whiskey: A Distiller's Journey Into the Flavor of Place, Rob Arnold

Epistenology: Wine as Experience, Nicola Perullo

Gastronativism: Food, Identity, Politics, Fabio Parasecoli

Printed and bound by CPI Group (UK) Ltd, Croydon, CR0 4YY
19/11/2023

08190872-0004